Patent Search
for Librarians and
Inventors

Timothy Lee Wherry

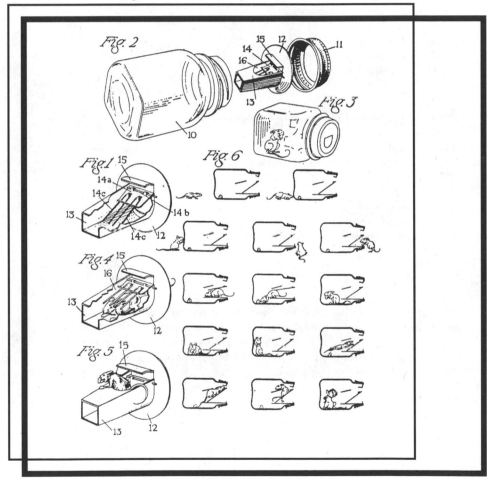

AMERICAN LIBRARY ASSOCIATION
CHICAGO AND LONDON, 1995

The paper used in this publication meets the minimum requirements of American National Standard for Information Sciences—Permanence of Paper for Printed Library Materials, ANSI Z39.48–1992. ∞

Cover designed by Richmond Jones

Composed by Publishing Services, Inc.,
   in New Caledonia on Xyvision/L330

Printed on 50-pound THOR paper,
   a pH-neutral stock, and bound in 10-point C1S
   by Malloy Lithographing, Inc.

Library of Congress Cataloging-in-Publication Data

Wherry, Timothy Lee.
   Patent searching for librarians and inventors / Timothy Lee
Wherry.
      p.   cm.
   Includes index.
   ISBN 0-8389-0641-9
   1. Patent searching.   I. Title.
T210.W44   1995
608.773—dc20   94-41416

While extensive effort has gone into ensuring the reliability of information appearing in this book, the publisher makes no warranty, express or implied, on the accuracy or reliability of the information, and does not assume and hereby disclaims any liability to any person for any loss or damage caused by errors or omissions in this publication.

Printed in the United States of America.

99 98 97 96 95   5 4 3 2 1

For Cindy

Contents

Introduction

Most people during their lifetimes are interested in patents. Usually the interest begins with an idea for a product that the person has invented or is thinking of inventing. Although most people, if they persevere, eventually seek out the services of a patent attorney, the starting point in the search for patent information is a library.

Librarians at large metropolitan public libraries receive requests for patent information daily. Even at smaller public libraries, patent questions are routine. Librarians in special libraries, academic libraries, and research centers also have to field a variety of questions dealing with patents. The problem is that most patent questions cannot be answered with a quick response and also, most librarians have not been trained in patent information. The patent system is confusing and complicated even for those librarians who work with it all the time.

One solution to this problem is that libraries hand out short instruction sheets to guide the novice in performing a patent search or in otherwise locating patent information; however, many people need something more detailed. Library users are referred to more detailed information in the book collection or in government documents that can give more patent information than is possible through a busy reference librarian or an instruction sheet. Unfortunately, a layperson or a librarian is likely to find most reading material about patents to be far too legal-oriented, complicated, and dull.

The state of available patent information for novice and librarian alike seems woeful. The majority of books on the topic are written in textbook form, generally by legal professionals, who direct their messages to other attorneys or to those who already possess a basic knowledge of patents.

Even recently published books on patent information begin with the assumption that the reader already knows something about patents. They take the reader on a journey through patent databases that can be searched only with the aid of a specially trained librarian. They present technical explanations, and detailed demonstrations of forms, laws, and esoteric trivia for which a novice has no need. Many librarians and amateur patent searchers have related to me that what they really need is a self-help book that will enable them to work on their own and get started quickly.

In over twelve years of presenting patent seminars, teaching patent classes, and assisting people in locating patent information, I have found that a conversational tone and simple straightforward information about patents was possible. People are fascinated by patents, but frustrated in finding some way to get sufficient information quickly, and presented at their level. There is a need for an interesting, easy-to-understand book about this interesting topic. It was from this basic premise that I decided to write this book. It is my intention that the reader enjoy this book and, more importantly, gain an adequate knowledge of patents so that the reader can make informed decisions about his or her creative efforts.

This book will not make the reader an expert, but it does provide enough information that a patent search or an understanding of the system can be undertaken without a great deal of confusion and toil. It will give laypersons and librarians enough information to be able to converse with legal professionals in an effective manner. In addition, the appendices give the reader valuable, basic information including a collection of often asked questions and their answers, a list of Patent and Trademark Depository Libraries across the country, and a list of Patent Documents in Federal Depository Libraries.

As the reader goes through this book it will become clear that patenting is not an exact science. Even today experts disagree in certain areas while other areas are unlikely to be disputed. I have always recommended that while an initial effort into patenting can be performed adequately by anybody with average intelligence, a person serious about pursuing a patent should work closely with an attorney who is registered to practice patent law.

The effort of writing this book was one that was undertaken by a single individual, but as with most large projects, assistance was given by many others. Some have helped in reading the manuscript and in offering suggestions to make the book better; some have encouraged my writing and published my short articles on patents; some have anxiously waited for me to write the next chapter so that they could offer suggestions (and have the information before anybody else); and some have offered encouragement and praise when it seemed the manuscript would never be done. I would like to thank Keven Harwell, Penn State Patent Librarian, for his suggestions and proofreading; the Patent and Trademark Depository Library Program for making recommendations, starting me on this road, and allowing me to speak twice at their conference; V. S. Dodd, editor-in-chief of *World Patent Information* for showing interest in this project; Mary Redmond, past editor of *Documents to the People,* for demonstrating a market exists for self-help patent information; Mary Claire Sprott of Sunnyvale Public Library and Larry Kueneman of Encinitas, California, for anxiously prodding me for the next chapter; Kjell Meling, director of academic affairs at Penn State Altoona Campus, for supporting my funding requests so that I could do the research necessary for this book; and my family, Cindy, B. J., and Sarah, who were sure from day one that I could do it.

Chapter One

The World of Patents

The right of individuals to hold patents is guaranteed by Article 1, Section 8 of the U.S. Constitution:

> Congress shall have the power . . . to promote the progress of science and useful arts, by securing for limited times to authors and inventors the exclusive rights to their respective writings and discoveries.

Each of the creative properties of patents, copyrights, and trademarks requires by its nature its own set of laws and methods of control. Together these laws and practices form the subject of intellectual property; that is, the Constitution grants the rights of patent, copyright, and trademark protection, but the manner in which each is secured, the time frame allowed for protection, and other details are detailed by the laws that were later written and included in Title 17 and Title 35 of the U.S. Code. This book is concerned with one aspect of intellectual property: patents.

The framers of the Constitution realized that the right to exclude others from the financial rewards of one's inventions would be an incentive to create and encourage business in the new nation. For the first time in the history of the world, a constitutional instrument recognized that individuals have a right to protect their intellectual property. Thomas Jefferson, the first patent commissioner, spoke of patents as "the locomotive that runs industry" and was aware of how important patents would become to the U.S. economy.

Most consumer goods that are offered on the open marketplace exist because the manufacturer was able to gain an advantage over the competition by prohibiting others from copying his or her unique product. This may seem contrary to the concept of a free market, but the advantage given by intellectual property protection actually helps, rather than hinders, the economy. With a patent, an inventor can control the design, manufacture, licensing, distribution, and copy of inventions. A patent is, in effect, a temporary legal monopoly over a specific device or process.

The first patent in the United States was granted by the state of Massachusetts in 1641, 135 years before the founding of the nation. It was granted to Samuel Winslow for a unique process for manufacturing salt. The first patent on a mechanical device in the western hemisphere was issued to Joseph Jenkes in 1646 for a water-propelled machine that was used to manufacture scythes. In this invention a water wheel powered a hammer, which beat metal into usable implements. Previously, scythes were manufactured by hand. The first U.S. patent was issued on July 31, 1790, to Samuel Hopkins for his new process of making potash, an ingredient in soap. In the first year of the U.S. system, Patent Commissioner Jefferson granted just three patents. Currently, about seventy thousand patents are issued annually. Jefferson himself was an inventor and developed a revolving chair and a stool that could fold into a walking stick, although he never patented the devices. One wag said that the revolving chair was invented so that Jefferson could look all ways at once.

Until July 13, 1836, when a classification system was devised, patents were arranged alphabetically by the inventor's name and kept in wooden boxes. There were 9,957 patents granted up to that time. In 1836, the first numbered patent was granted to J. Ruggles for notched train wheels, which permitted better traction on uphill grades.

As the first patent searcher, Jefferson insisted that patents be arranged in such a way that they could be researched easily. Jefferson kept drawings of his inventions in old wooden shoe boxes, and this became the standard filing system. A visitor to the Patent and Trademark Office (PTO) in Washington, D.C., can still see the millions of U.S. patents shelved in metal and wooden "shoes." (Actually the PTO is located in Crystal City, an office and commercial center across the Potomac in Arlington, Virginia, but it is commonly referred to as being in the District of Columbia.)

Few people are aware that during the Civil War the Confederacy had its own patent office, headed by former Union patent examiner Rufus Rhodes. The Confederacy's first patent was issued on August 1, 1861, to James Houten for a breech-loading gun, and the last Confederate patent, number 266, was issued on December 17, 1864, to W. Smith for a percussion cap rammer. Understandingly, many of the Confederate patents were for war machines. The Confederate Patent Office was destroyed during the battle of Richmond in April 1865, and practically all of the models and records were lost. Some still exist in scattered private collections.

The PTO is an anachronism in the twentieth century in that virtually the entire patenting process, from application to searching for existing patents, is performed manually. Not only do over five million patents have to be categorized and shelved, but approximately 165,000 patent applications are submitted annually, and hundreds of thousands of accompanying documents must be controlled. It is estimated that the PTO has a total of thirty million documents relating to patents, and, according to the *Wall Street Journal*, at any given time nearly two million cannot be located. It is important that the PTO retain all of these records; about 65 percent or seven in every eleven applications submitted to the PTO are eventually granted a patent, and accurate records are indispensible in case of an infringement suit. Patent examiners have reported stories of a colleague who had run out of room in his office and stuffed so many patent applications in the ceiling above the soundproofing that the ceiling collapsed.

The government is willing to subsidize education for examiners and many attend law school at night. Since the highest pay rate for an examiner is about $80,000 a year and patent attorneys can easily make $150,000 a year or more, many examiners let the government pay for their law degree and then leave the PTO to practice patent law.

In recent years the PTO has embarked on a $1 billion project to automate the patenting system. Some searching procedures have been automated and private vendors have made computerized databases of patents available. A description of CASSIS, an automated system, can be found in chapter 6, and additional online services are listed in this book's bibliography. To have an efficient system, the PTO may have to rethink 150 years of procedures that involve pushing paper from one location to the next. What is most important in automating the system is to successfully integrate text and drawings in one file that may be searched by computer. At this writing, the PTO claims to be close to having such a product.

Successful Patents

An indication that the old manual filing system worked—at least for simple inventions—is that over the past two hundred years no other country can show a better record of innovation. There is still a belief that a patent on a simple invention will make an individual rich and secure for the rest of his or her life. As Emerson reportedly said, build a better mousetrap and the world will beat a path to your door.

Sometimes this is true. In 1874, Joseph Glidden was granted patent 157,124 on barbed wire (see Drawing 1–1). This was a revolutionary device that for the first time permitted ranchers on the treeless plains to divide ranch land effectively and economically. By 1887, 173,000 tons of barbed wire were being sold annually. Another patent success story shows that in 1948 a Swiss engineer, George de Mestral, was annoyed by the tenacious grip of cockleburs on his socks after a hike. By looking at the burs under a microscope he discovered that they were shaped like hundreds of tiny hooks that attached to the threads of his socks. De Mestral got a weaver in Lyon, France, to make a facsimile of the burs by hand. De Mestral improved on his original design, which he called "locking tape." Patents were secured worldwide on "velvet type fabric and method of producing same." (His U.S. patent, 2,717,437, was granted in 1955. See Drawing 1–2.) By the late 1950s, de Mestral had trademarked the name Velcro, and looms were turning out sixty million yards of Velcro annually.

Of course, not all patents are simple. In 1942, Chester Carlson was granted patent 2,297,691 (see Drawing 1–3) on something called electrophotography, a process that copied documents without the use of inks or liquids. An important point to remember is that a patent need only exist on paper; a model or prototype has not been required since the nineteenth century. Carlson had patented a process, not the actual machinery that would perform the process. Unable to convince anybody of the viability of the process, he assigned the patent rights in 1944 to the Battelle Memorial Institute in hopes that they would build an electrophotography machine. Unable to do anything with the

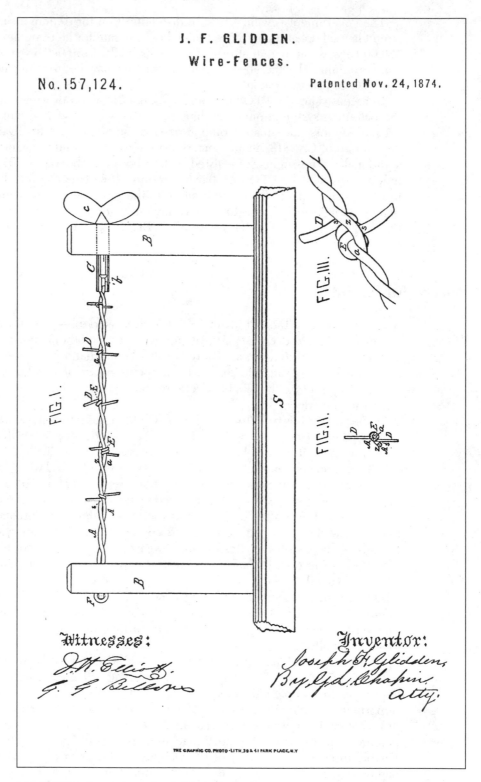

DRAWING 1–1.   Joseph Glidden—barbed wire

Sept. 13, 1955   G. DE MESTRAL   2,717,437

VELVET TYPE FABRIC AND METHOD OF PRODUCING SAME

Filed Oct. 15, 1952

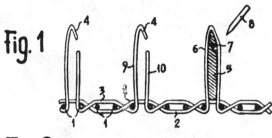

Fig. 1

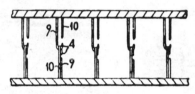

Fig. 2

INVENTOR

George de Mestral.

BY

ATTORNEY

DRAWING 1–2. George de Mestral—Velcro

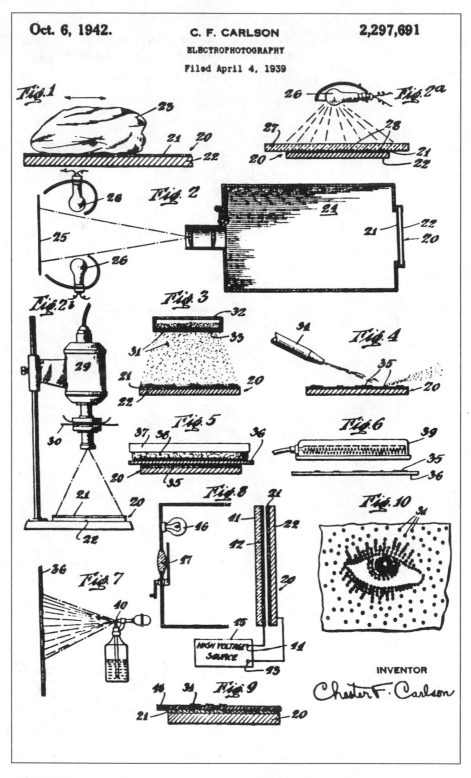

DRAWING 1–3.   Chester Carlson—electrophotography

patent, Battelle offered an option to the Haloid Paper Company. Haloid felt that if they could develop a machine from Carlson's patent, they could sell more paper. Haloid developed the machine during the next thirteen years, changed its name to Xerox, and sold the first photocopy machine in 1962.

Since an invention has to exist only on paper, some very interesting applications have been submitted. The most interesting was an apparatus for keeping a severed human head alive. The PTO initially approved the "Device for Perfusing an Animal Head" application, which was supported by four sketches and eleven pages of medical double-talk. It was later discovered that a patent lawyer sent it in as a joke.

Many companies obtain patents to control the technology of an entire industry and never intend to produce the product. Jerome Lemelson, a lone inventor who obtained patents in the 1950s on everything from toys to computer vision technology, has recently earned over $200 million by threatening to sue automobile manufacturers for infringement on his patents. Lemelson's cluster of patents on "machine vision," filed in 1954, outline methods, now routine in factories, for using robots and bar coding scans. Japan's twelve leading car companies alone have paid more than $100 million to Lemelson to settle out of court.

Lemelson is also significant in that he is said to hold the PTO record for a "submarine" patent. (*New York Times,* February 21, 1994, B2.) A submarine patent is one that has been slowed down, or submerged, for years in its application procedure. To slow down the patenting procedure, some inventors deliberately make obvious minor mistakes in the wording or in the drawings. When the patent examiners discover these errors, the patent application is returned to the inventor, then corrected and resubmitted. In the resubmitted patent, another error is discovered and so on. Each error takes many weeks, sometimes months, to correct.

By drawing out the examination procedure, an inventor can effectively prohibit others from patenting the same invention. The advantage is that anybody who makes a product based on the submarine application will owe the inventor licensing fees once the patent is granted.

Lemelson applied for his record submarine patent on "apparatus and methods for automated analysis" more than forty years ago. When the process is delayed, in effect, the length of time allowed for patent protection is extended into a more profitable time period. To prohibit this practice, Congress is now holding hearings on a proposal to limit patent protection to twenty years after the application is filed rather than seventeen years after the patent is granted.

The classic case of industry control by patent ownership is that of the Selden auto patent. George Selden was a bright patent attorney who was also an amateur inventor. Fascinated with the concept of a self-propelled vehicle, in the late 1870s he developed and built a gasoline engine attached to a carriage. In his invention were the basic components of automobiles: a simple two-cycle engine, carriage, clutch, and running gear. It was hardly a marketable or practical vehicle, but on May 8, 1879, he applied for a patent (see Drawing 1–4).

Since Selden was a patent attorney and knew the status of patents under consideration at the PTO, he knew that his was the first patent application on an automobile. Selden managed to stall the patenting procedure by taking advantage of quirks in the system and was not granted patent 549,160 until

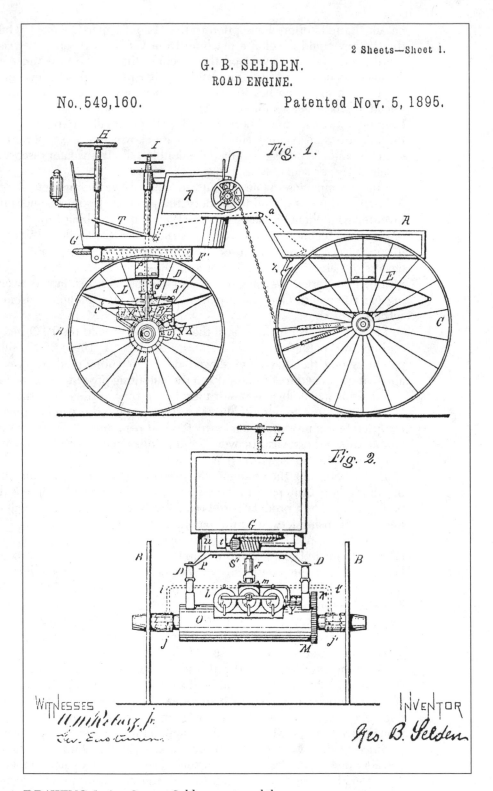

DRAWING 1–4.   George Selden—automobile

November 5, 1895—sixteen years after his application. During this time it was also possible to amend an application, which Selden did until his patent included virtually every major component of the automobile. It did not matter that during this time a superior four-stroke engine was developed or that Germany's Daimler was widely acclaimed as the true inventor of the practical motorcar; Selden was first in line. His patent would run seventeen years until 1912.

In 1899, fearing suit by Selden, a group of auto manufacturers approached Selden for a license since they could not produce any significant part of an automobile without infringing on Selden's patent. Selden approved and sat back and waited for the royalties to arrive. He made $200,000 because of his patent. He was challenged only by Henry Ford, who endured an eight-year court battle. Ford finally won the suit on appeal and broke Selden's monopoly in 1911, one year before the patent would have expired.

Edwin Land, inventor of the Polaroid camera, held patents on every component of the instant camera and the process of instant development, prohibiting infringers from producing even minor subassemblies of his invention. Only when Polaroid improved the camera or the film and introduced a new model would it allow others to license its technology, but Polaroid only allowed older, obsolete models of an instant camera to be licensed to others. Kodak, which lost a $909.5 million settlement to Polaroid in the early 1980s, was found to be infringing on the process of instant photography by producing its own instant camera and film and was required to reimburse purchasers of their film and cameras with cash and coupons for other Kodak products.

A brilliant invention, however, may be useless without the technology to make it practical. Bell's telephone (patent 174,465, see Drawing 1–5) required the additional inventions of switching devices, amplifiers, transformers, and transmission mechanisms. Once certain large patents are granted, a new company may need to be started to obtain the capital support, management style, and technical expertise to make a truly great invention successful. Bell's invention, for example, gave birth to the Bell Telephone Company, but no better example illustrates this need for ancillary equipment and business acumen than Thomas Edison's electric light.

Edison did not invent the electric light in 1880. It was invented as early as 1838 when a French scientist passed an electric current through a carbon rod sealed inside a vacuum chamber and watched it glow. In 1878, two years before Edison was granted patent 223,898 (see Drawing 1–6), Joseph Swan demonstrated a carbon filament vacuum lamp in England but did not patent it. Evidence suggests that Edison knew of Swan's discovery, since his Menlo Park library rivaled university research libraries of the time. Edison obtained a U.S. patent on the electric lamp and when Swan attempted to put the electric light into production in the United States, Edison blocked it with litigation charging that he owned the patent. Both men soon realized that the invention was useless without an electrical power source. Nobody would buy an electric light if there was no place in their homes to plug it in. Edison and Swan dropped their legal action, joined forces, and formed the Edison & Swan Electric Company in 1883. By 1910, three million American homes were using electric light bulbs.

But a simple *or* complex invention is only rarely successful. Only about 65 percent of patent applications ever become patents; only about 10 percent are

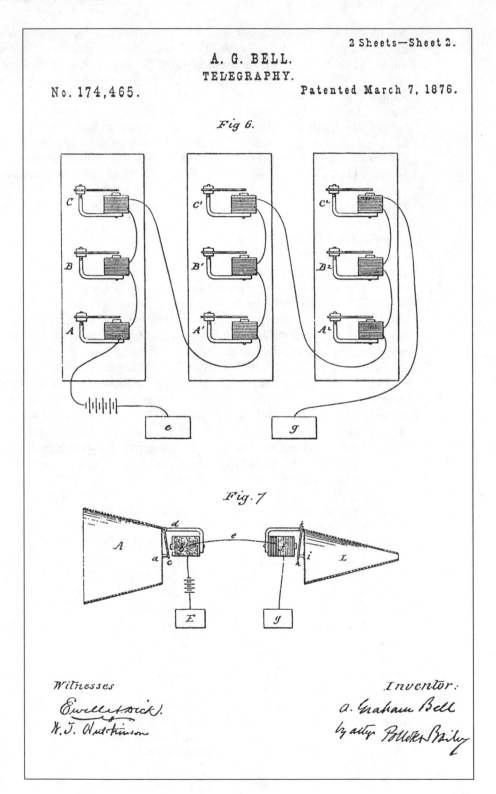

DRAWING 1–5.   Alexander Bell—telephone

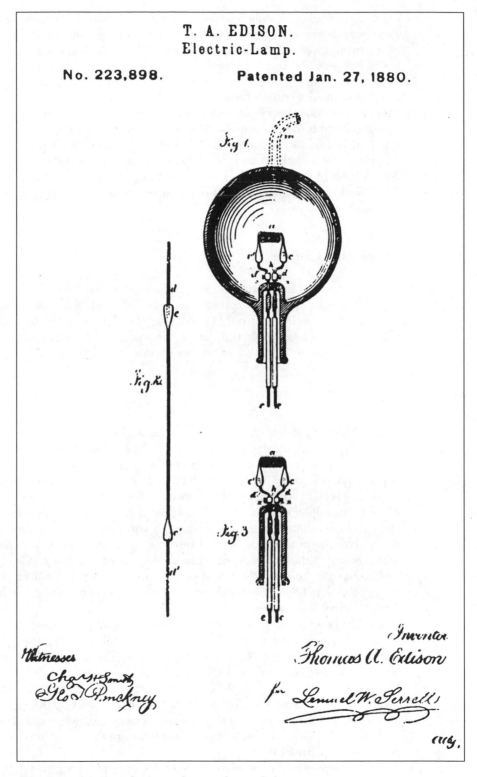

DRAWING 1–6.   Thomas Edison—electric light

eventually marketed; only 3 percent of patents issued ever make a dime for their inventors; and the cost of obtaining a patent typically is outside the reach of a private inventor. Even if an inventor files a patent application *pro se*, on his or her own, the cost is generally several thousand dollars. Patent law is designed to encourage individual effort in that only individuals, not corporations, may be granted a patent; however, large companies with research and development divisions simply pay patenting fees and have the employee inventor assign the rights to the company. The process of selling a patent or of licensing a patent is a relatively simple matter and is discussed in chapter 2.

The above anecdotes and examples demonstrate several of the important foundations of the system. But before wading too deep into patenting, librarians should be familiar with the basic concepts and terms of patent law and processes.

Patent Terms and Concepts

A patent is a contract between society and inventors. In the interest of spurring innovation, society agrees to protect an inventor's control over an invention. In return, the inventor must publicly disclose the details of the invention. The information made available thus potentially spurs further innovation. A patent protects the *implementation* of an invention's underlying idea.

Patent Types

Three types of patents exist: utility, design, and plant.

Utility patents are what most people think of as an invention; for example, a machine or a process. Utility patents are further broken down into three types: chemical, mechanical, and electrical. Some patents are squeezed into these three categories when their placement is not obvious. Computer software, for example, in some instances is patentable as a set of instructions to an electrical component, thereby making the patent fall into the electrical category. Pharmaceuticals are placed into the chemical category. (The United States is one of the few countries that allows patents on drugs.) Gene splicing techniques or the recently patented Harvard Mouse, a strain of laboratory mouse that has been genetically manipulated to contract cancer readily, fall into the chemical category.

Design patents are granted on the appearance of something. In making a chair with a pedestal instead of four legs, for example, an inventor does not re-invent the chair, but only changes its appearance. The design can be protected, not the concept of the chair itself.

Copyright, although thought of as protection of written works, also offers protection on designs. Copyright may be granted on jewelry designs, artwork, and sculpture, for example. The difference between a design patent and copyright is that a design patent is granted for a design that is on or part of a utilitarian article of manufacture.

A sculpture can be granted a copyright, for example, because it performs no useful function in the practical sense; it is simply nice to look at. A sculpture that has had holes drilled in the top and bottom and wired as a lamp would receive a design patent because the sculpture is part of a functional item, a lamp.

Plant patents are granted on bushes, trees, roses, and so on that are reproduced asexually. Asexually propagated plants are those reproduced by means other than from seeds, cuttings, tuber propagation, or graftings. Patent roses, for example, are a particular type or color of rose that is protected from duplication by other horticulturalists. Interestingly, since a patentable part of a plant is its color, these patents are issued in color booklets rather than the standard black-and-white patent document. A problem for librarians is that microfilm of plant patents requires a color reader/printer since copies of the patent must be in color.

Copyright

Copyright is protection for the *expression* of an idea. Even legal professionals are sometimes confused over the differences between a patent and a copyright. Copyrights apply to books or other written work, pieces of art, sculpture, computer programs, architectural works, jewelry designs, music, and maps, for example. One would patent a mousetrap, but copyright an article about a mousetrap.

With copyright, expression is everything, utility is nothing. The important difference between copyright and function is the concept of *function:* the thing being protected must be useful in order to be patented. One would copyright a book because it is useless, in a utilitarian sense. One would patent a doorstop that was shaped like a book because it performs the function of holding open a door.

Although computer programs are usually copyrighted as written works, some critics maintain that software, like other nonsoftware inventions, cannot be copyrighted because it performs a function; however, if the software represents only one of several ways of performing a task, it is not unique and will be denied a patent. The distinction to the PTO between copyrights and patents for software is in how the program uses algorithms, a set of instructions that tells a computer what to do. Software can be patented only if it is the sole algorithm for performing a function; mathematical algorithms are considered laws of nature and as such are not patentable, but computer algorithms are patentable. Because of these and other issues affected by operating systems, the distinction between patents and copyright is still fuzzy. On certain types of software, the Patent and Trademark Office grants only about two hundred patents annually.

Trademark

Trademark is protection on slogans, logos, or product names. The primary function of a trademark is to indicate the origin of goods and to distinguish them from those sold or manufactured by others. A person selling a quality product wants to advertise in some way that his or her goods are superior to another's and to create and maintain a demand for the product. Unlike other forms of intellectual property, a trademark has to be used first before it is granted protection. Not only does the trademark have to be used to qualify for federal protection, it must appear on goods sold across state lines. A trademark must be used on goods first to prohibit companies or individuals from simply registering every trademark they could think of in an effort to control the creative atmosphere of a given industry. Of course this still happens to some extent, but it is more difficult if the goods must first carry the trademark and be sold before rights are granted. More about trademarks and copyrights appears in chapter 5.

Requirements for Patenting

A U.S. patent has three requirements.

1. It must be the first of its kind. The patent statutes mention a concept called "anticipation." This means that the invention may not have been disclosed before an application is submitted. The rationale for this rule follows the whole basis behind patenting in the United States as opposed to patenting in some other industrialized nations: In the United States the person who first invents or "reduces to practice" a device has the right to patent; in some other countries, Japan for example, the first person to obtain the patent has the right to the invention. To protect one's patent right in the United States, then, it is necessary to assure that the idea is an original thought of the inventor by verifying that the idea did not exist before application was made. There must be no other patent like it not only in the United States, but in the world. Also, there must be no mention of the invention in any printed matter, such as a magazine or newspaper, or any way to show that the invention was in public use. In one case an invention was considered to be public knowledge because it was presented in a single copy of a thesis in Russian that was held by a Chinese library.

2. It must be useful. This means, as pointed out above, that a patent must have a *function*. This is rarely a problem. If an invention is not useful, an inventor is unlikely to go to the time and expense of obtaining a patent. Of course, usefulness is sometimes a questionable call. "Rube Goldberg" devices, complicated machines that perform simple tasks, are patentable and useful but applying for a patent on such devices is probably not worth the time and expense to an inventor. Chapter 2 of this book contains a discussion of frivolous patents. The only exception to the concept of usefulness would be in the case of plant patents where the patent is simply attractive and performs no function.

3. It must not be obvious to others of ordinary skill in the field to which the patent pertains. This is the ultimate condition, and it attempts to measure an abstract thing: the technical accomplishment reflected in the invention. For example, substituting one material for another or changes in size are usually not patentable. However, an examiner must weigh the minor changes by which a new invention accomplishes the same end as an existing patented invention. This sort of judgment calls for a technical background.

Patent Examiners

Some of the best-educated bureaucrats are at the PTO in Washington, D.C. Many hold doctorates in their field and have legal training. These are the people who actually look at patent applications to determine if an invention is new, useful, and not obvious. The PTO employs about 1,600 patent examiners and 200 trademark examiners. More on examiners appears in chapter 4.

Patent Pending

Often seen on new products, this term means that a patent application has been submitted and is simply a marketing technique to give the manufacturer a head start. Since the inventor of the product is first in line for a possible patent, others are discouraged from applying for a patent on the same product. The term *patent pending* offers no protection under the law since an application may be rejected, but it is unlawful to mark an item as such if an application is not on file. A person may copy and manufacture an invention marked patent pending without fear of litigation until the first-in-line patent is issued. It is unlikely that a person would steal an idea marked patent pending, however. Once the pending patent is issued, the copier would have to shut down production, and the time and expense devoted to manufacture would be lost.

Also, during the application procedure, the PTO keeps patent forms secret so that no others may copy the device. Employees of the PTO are prohibited from owning patents to further assure the confidentiality of the process.

Patent Expiration

A patent is good for seventeen years from the date of issue; design patents are valid for fourteen years. A patent cannot be renewed or extended except by an act of Congress. To keep a patent in effect, the inventor pays maintenance fees for those seventeen years. Once a patent has been granted, even though it has expired, the same device cannot be patented again by the original inventor or by another.

How then has a product such as Coca-Cola, invented in the nineteenth century, had its soft-drink formula protected? After all, since after seventeen years the patent became public property, anyone could copy or manufacture it. Answer: The formula is not patented; it is a trade secret and need not be released publicly. Chapter 5 has more on trade secrets.

Another question librarians are often asked about patent expiration: If the formula for Tide laundry detergent expired after seventeen years and everybody can copy the formula, how does Tide keep its customers happy? Isn't the same detergent available from other manufacturers at a cheaper price? The answer is that the company comes up with a new, improved Tide, which has an altered chemical formula and is granted a new patent. Who then would want to buy the old, unimproved Tide? Of course, Procter & Gamble could try to keep Tide's formula a trade secret, but that is difficult to do in the detergent industry, where formulas are easily analyzed by chemists.

The same concept holds for drug manufacturers. A successful drug such as Valium can make a lot of money for a pharmaceutical company, but after seventeen years the company is pressured to come up with a new tranquilizer since other drug companies may then manufacture Valium as a generic drug. Generic drugs are available from the pharmacy because their patent has expired.

Patentee and Assignee

A patentee is the person who is named as the inventor of the patent. An assignee is the person or company to whom the patentee has given rights to the invention. A patentee may also license or sell a patent. Licensing patents permit others to manufacture, distribute, or otherwise use an invention by paying fees to the patentee while the patentee maintains ownership of the patent. In selling a patent, the patentee relinquishes rights to the patent for an agreed payment.

When a patentee licenses, assigns, or sells a patent, it is done for a variety of reasons: the inventor needs money immediately, the inventor cannot afford to manufacture the invention, the inventor has no knowledge of marketing techniques, and so on. In some cases employees' contracts state that they must assign rights to any patent obtained during the time of their employment to their employer.

The patentee does not have to be a single individual, it may be several people holding joint rights if they all contributed intellectually to the invention. "Contributed intellectually" means that they were involved in the actual inventing of the device. For example, if a brother gives an inventor $1,000 to finance the development of a widget but has no input into the actual inventing of the widget, the patent cannot be issued in both the inventor and brother's names; however, a legal contract may be made in which the brother may be a creditor to whom a percentage of the invention's profits are paid. The brother may be the assignee if he has filed forms with the PTO showing that he has purchased rights to the invention, but only the true inventor(s) can actually obtain a patent.

In many cases, large corporations make the right to become assignees to an employee's patent a condition of employment. Any employee in these situations is legally bound to transfer rights to the employer for an invention related to the company's work that was developed during hours of employment with the company.

After the PTO grants a patent, it does not act in an inventor's behalf except in cases of reissues or re-examinations (see below). Any violation of an inventor's patent rights is pursued through the courts and not through the PTO. Many inventors have been ruined economically by fighting infringement cases in the courts and some patentees attempt to sell their patents if litigation is pending, since it then becomes the assignee's responsibility to defend the patent in court.

Reissued Patents

When patents are issued, the patent document contains many pieces of information: the name(s) of the inventor(s), a brief description of the invention, several drawings of the invention from differing angles including close-ups of components or subassemblies, "claims," which are the words that explain precisely what is new and unique about this invention, a long description of the invention and sometimes a brief background of the technology or need that necessitated it, and a short list of related patents and classifications searched. On occasion, a patent is issued that contains a mistake in one of these important components. There may be a mistake in the wording of a claim, for example. When this happens the inventor fixes the problems and a new number is assigned. In lists of patent numbers that a searcher may use, these reissued patents are easily identified by an RE before the number.

The first part of this book has attempted to introduce librarians to basic patent history, terminology, and procedures. The next chapter will try to clear up some of the misinformation that surrounds patenting.

Chapter Two

Misconceptions and Myths about Patents

In the two hundred years that the patent system has been in use, misconceptions and myths about patents and patenting have established themselves in the minds of the public. Librarians are often confronted with this false information when assisting patent searchers or inventors. It is important that librarians have a more realistic view.

This chapter has been written with amateur inventors in mind, since it is they who are most likely to seek assistance from librarians. Librarians should also be aware that in debunking the misconceptions held by many amateur inventors or patent searchers, one can discourage patent research. Be tactful.

Risks and Rewards

Patenting an invention does not lead to easy wealth. Most often the opposite is true. Until the twentieth century the United States was a rural nation and large-scale economic development still lay ahead. Many individuals had learned to be self-sufficient and many were self-educated. In this environment, individuals with technical talent could arise and become rich on the invention of a single simple device. Today, however, this is rarely the case.

Only about 2 percent of inventions are economically successful for a variety of reasons: an inventor may have no way of producing his patent; he may have underestimated the demand for a new product; the ancillary technology to support extensive consumer use of the invention may not be available (as in the case of Edison and Bell mentioned in chapter 1); or the invention may be infringed upon without the true inventor's having the economic resources to pursue court action.

The belief that if a person builds a better mousetrap (see Drawing 2–1) the world will beat a path to the inventor's door is a false belief in today's world. More likely the inventor must beat a path to the door of the world—that is, aggressively market and promote the invention. Although Victor Hugo wrote, "Greater than the tread of mighty armies is an idea whose time has come," the

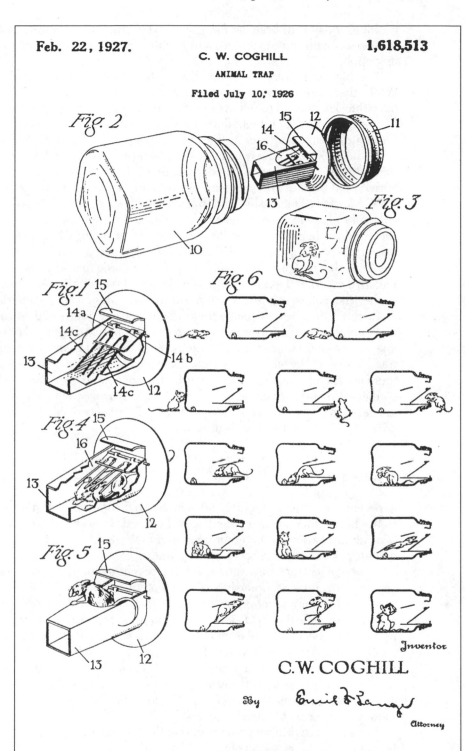

DRAWING 2–1.   C. W. Coghill—animal trap

problem today is in bridging the gap between the person with creativity and the person with the capabilities and facilities to market and manufacture the invention.

The biggest problem inventors have is licensing or selling their invention. While there are many dreamers, creators, and innovators, few of them also have the skill needed to sell. As mentioned before, many great ideas never are marketed and never make a dime for their inventor. One case of an inventor finding the right marketer at the right time is that of Emma DeSarro.

In the 1950s, a young woman, Emma DeSarro, created a design for a roller skate. It was no ordinary roller skate in that the wheels were in-line. Ms. De-Sarro spent $900 of her own money developing a prototype and getting patent 3,387,852 for "Detachable and Removable Roller Skates" based upon the concept of "single longitudinal wheels mounted on special frames." But when she tried on her own to get the Chicago Roller Skate Company to buy the patent, she was rejected. In desperation she contacted an invention marketing company that took her money, but provided no marketing services. Undaunted, DeSarro continued to improve on her skate and patented a new design so that as the first patent expired, a patent on the new design was granted.

In the 1980s, several marketing people became interested in the skates, most notably Mary Horwath. Horwath saw the potential of marketing the in-line skates to urban youths not attracted by traditional roller skating. With the trade name Rollerblades, Horwath licensed DeSarro's patent and marketed them as speedy footwear for hip urban teenagers. The Rollerblade carton contained pictures of city graffiti, and the advertising campaign conveyed an image of recklessness and independence. The success of Rollerblades has created a $250-million-a-year industry.

Assigning and Licensing Patents

A patent is personal property and may be sold, bequeathed in a will, passed to the heirs of a deceased patentee, or licensed. In addition, assigning and licensing are two ways to get an invention marketed and manufactured if the inventor decides not to take on these duties.

To assign a patent, an inventor sells the patent to another party; in licensing, an inventor retains ownership of the patent. In selling an invention, the patentee has only to agree on a price with a buyer and file a Certificate of Assignment with the PTO along with a nominal fee. The party who buys the patent then becomes the owner, with rights to manufacture, distribute, and control the invention. However, the agreement should be handled by an attorney since there are several factors to take into consideration. The price should be high enough to reflect the time, effort, and money spent on the idea as well as its earning potential, but low enough to realize the risk and development money a buyer would need to invest.

Licensing is a different matter in that the patentee retains ownership of the invention, but allows others to market and manufacture it. Usually an inventor will agree on a percentage of profits when a patent is assigned. Again, any such agreement is best left to qualified attorneys, not only to handle the agreement but also to advise the inventor on the best course of action.

The issues to be decided by an inventor in licensing an invention are more complex than simply deciding to whom to license it. For example, is the licensee required to produce a certain number of units per year? If not, the licensee may simply purchase the rights and sit on the invention so that the patentee receives only an advance paid when the agreement was signed and does not receive a share of the profits. Does the licensee have exclusive rights, or does the patentee want to license the invention to more than one party? This may be a consideration when one assignee's marketing ability is not national or international or when a licensee markets to only one type of industry. There are other considerations also: the ability to audit the assignee's financial records, the assignee's responsibility in case of infringement, the handling of the invention in case of bankruptcy of either party, and the right to terminate the agreement. As said, an attorney is best qualified to answer these questions.

Court of Appeals

Before 1982, if a true inventor and holder of a patent took legal action against another who copied that patent without buying or licensing, the patent holder had only a 50 percent chance of winning the suit. In 1982, a new Court of Appeals for the Federal Circuit, a specialized appellate court in Washington, was formed to deal exclusively with patent cases. This court not only hears patent cases exclusively, but also has taken a firmer stand on protecting the rights of the patent holder against infringement. Since that time, patent settlements in favor of the true patent holder have increased significantly and the power of patent ownership has been strengthened, as in the case of Jerome Lemelson mentioned in chapter 1. At the same time, the new court has reeled in other cases it deemed frivolous, cases that previously had languished for years in lower courts before a decision was reached. For example, Mr. Lemelson lost a case involving his claim on the patent of Mattel's Hot Wheels toys because claims that he had previously invented a miniature toy car were judged frivolous.

Withholding Manufacture

Many inventors believe that if they invent a product that will cut into the profits of a large corporation, the corporation will buy the patent rights to keep the new invention off the market. One particular story about a carburetor that will allow an automobile to achieve more than one hundred miles per gallon shows up every five years or so. In reality, the modern automotive marketplace is so competitive that any advantage would be promoted and accepted as a means of selling more product. Curtailing competition in the worldwide automobile industry is an absurd notion. However, that being said, some companies in other industries do buy or attempt to control the patents of an entire industry to stop the manufacture of competing products.

For example, the in-line skating industry is growing rapidly. Patents on new developments in in-line skating have several more years to run and manufacturers have large inventories of skates based on these existing patents. If a new, improved in-line skate is patented, it is in the best interest of a company to purchase the patent and sit on it until the inventories of older skates are depleted. If a single manufacturer can control all the new developments in a particular industry, it can effectively control the industry and stifle competition.

A patent search, explained in later chapters of this book, often reveals that the unique idea that would lead to riches is already patented. One amateur inventor spent more than $1,000 on drawings, specifications, and a market study of a water shut-off system before deciding to do a patent search to see if a similar device was already invented. This inventor's idea was to place a valve and a solenoid on the water supply line to a house. The configuration was such that if the water pipe suddenly burst from a frozen pipe or other break in the line, the measured increase in water flow would trip the solenoid and close the valve, preventing water damage to the house. The flash of inspiration came to this inventor when the hose on his washing machine burst and flooded his entire house while he was away.

A simple patent search that took only one day and was done *after* the inventor had already invested a lot of money in preparing his patent revealed that such a device was already patented by a person who lived in Leadville, Colorado (a very cold place in the winter). The device was never successful because a valid market study revealed that water pipes do not burst often except in the case of the water line attached to a washing machine. That problem is easily solved by turning off the supply valve to the washer until laundry day or purchasing a hose enmeshed in a stainless steel fabric which prohibits the hose from bursting. Consumers regarded the expense of installing a shut-off device on a main water line as greater than the risk of flooding.

International Patents

The first thing that librarians should know about international or "world" patents is that there is no such thing. A U.S. patent protects the intellectual property of the patentee only in the United States. A U.S. patent is not enforceable outside the United States, nor is a foreign patent enforceable in the United States. Protection in each country is accomplished by filing an application in each country for which protection is desired. Widespread reports in the popular press tell of inventors who think that international patents can be obtained in one stroke. In reality, an inventor is usually advised to apply for a patent in each country in which protection is desired—an expensive and time-consuming procedure.

An alternative approach is to file an International Application under the Patent Cooperation Treaty (PCT). Again, it must be stressed that this does not provide an international patent, but minimizes the cost and procedures resulting from international patent filings. The PCT, signed in 1970, has forty-five participating countries and allows a patentee to file an international *application*. The application then has to be filed with each national patent office in which protection is requested. The advantage is that a single application may be copied and then submitted to each PCT country rather than having to produce a separate application for each country.

Under the PCT an international patent search is then carried out by one of the major patent offices, resulting in a report which lists documents in any

PCT participating country that may affect the patentability of the invention in a particular country. The negative side of the PCT is that once the application is filed there is little saving of time or expense as the inventor goes through each nation's patent procedure. Another negative is that since so many countries and patents are involved in the search, there is a greater chance for error than in a country-by-country search.

The PCT is not the only treaty governing international patents: for example, the World Intellectual Property Organization (WIPO) governs the Paris Convention. The Paris Convention, signed by eleven countries in 1883 and now signed by one hundred countries including the United States, has proved to be one of the most enduring and successful acts of international cooperation ever created. Although revised several times, the basic concept of the Paris Convention remains the same. An inventor who files for patent protection in any of the participating countries has twelve months to apply for protection in any of the other countries. These other applications filed within that time frame have the effect of being filed on the same day as the original application; that is, only the original inventor may file an application in any Paris Convention country for one year. Other applications, even if filed earlier in a participating country, are considered secondary to the original application.

The great advantage is that an application does not have to be filed in all countries at the same time. An inventor has a year to decide in which countries he or she may desire protection and to estimate the expense of preparing applications for each country. Despite enduring for more than one hundred years, the Paris Convention is now under attack by (so-called) Third World countries. These countries are appealing to GATT, the General Agreement on Tariffs and Trade, to make sweeping changes in the Paris Convention to provide open markets and free trade based on fair competition. These countries regard patents as constituting unfair competition, especially in the area of pharmaceuticals. In most of the industrialized world, pharmaceuticals are not patentable since their distribution serves the good of all mankind; however, some countries, such as India, see the ability to monopolize medicines as basic to their economic development. (As mentioned before, the United States is one of the few countries that allows patents on pharmaceuticals.)

WIPO, mentioned above, does more than serve as administrator for the Paris Convention. WIPO administers seventeen separate intellectual property agreements worldwide, but its biggest contribution is the creation of an international patent classification system.

Models

One recurring image about patents is supported by cartoons showing a person sitting outside a patent office door with a model of a ridiculous invention. The requirement of a model with each patent application was dropped in 1870, although for about a decade thereafter the PTO called for them frequently. The PTO still has a right to call for a model if necessary to prove operability, but seldom asks for one. In a recent application for the persistent attempt at a perpetual motion machine—a machine that produces more energy than it consumes—the PTO required a working model to demonstrate that such a

STEP TWO: Searching the *Index to the U.S. Patent Classification*

This paperback annual is held by most Government Documents Depository Libraries and by all Patent and Trademark Depository Libraries. A searcher needs it to locate the patent subject areas, called classifications, that categorize an invention. The searcher must have these classifications before proceeding with a patent search.

The *Index* is published each December, and it is important that a searcher use the most current edition. The patent system is a continually changing system, and an inaccurate classification found at this step can lead to a wild goose chase that will be both frustrating and time-consuming.

By looking up the terms identified in Step 1, which in the *Index* are called keywords, a searcher can get the classification of the invention being searched. The *Index* contains keywords arranged in alphabetical order (see Figure 3–1). Beneath each word is a more specific description. To the right of each word are two columns of numbers. The first is the major patent classification and the second is a subclassification.

For clarification of this step, an invention called a "fly mask" will be used as an example. A fly mask is a piece of cloth, usually canvas, that covers the head of a horse or a cow, and lets the animal eat and see while protecting it from flies.

In Step 1, a searcher would have broken down this device into its parts and listed any synonyms that might also describe the device. The components and synonyms might be: mask, face covering, flies, horse, cow, insects, canvas, and livestock. The goal is to look up in the *Index* as many synonyms and related words as possible to give as many access points to the *Index* as possible. It is better to have too many terms than to not have enough, since some terms will not appear in the *Index*.

Listed for "Mask" in the *Index* is the following:

MASK (See Face Guards)		
Baseball	2	9
False faces	2	206
Fire fighting	D2	7+
Gas mask	425	815*
Head covering with	2	173
Photographic	430	5
Respirators	128	206.12+
Surgical or oxygen	D29	7+
Swimming	D29	8+

FIGURE 3–1.   Listing for Mask

Much information is contained in this short listing. The "See Face Guards" reference in parentheses means that a separate listing appears in the *Index*, alphabetically under "Face Guards." The indented terms list the various types of masks that are classified in the system.

The invention being searched, a fly mask, is not listed here, but this is not a dead end. The "see" reference gives the searcher another category. The searcher would refer to Face Guards in the *Index*.

The Symbols

Before looking further, the searcher should learn the meanings of the symbols that accompany the numbers in the two columns to the right. Knowing their function will save time and make the process easier to interpret.

Patent classes are numbered rather than described with words. The "D" in front of some of these class numbers means that this type of patent is a design patent. When most people think about a patent they think about a utility patent, which is only one of three types of patents. Remember, in addition to utility patents, there are plant patents and design patents. Plant patents are granted to unique strains of plants, like rosebushes. Many people are familiar with patent rosebushes. A design patent is protection on the design of something rather than protection of the device itself. For example, a chair may be supported by four legs, a pedestal, or metal tubing, or it may be shaped in some unusual way. In securing protection for these variations, a chair designer does not reinvent the chair, since the chair's function has not changed, but makes the chair appear differently. That is, the chair itself cannot be patented, but the critical feature of design can be patented. This design would be protected by a design patent. If the searcher is not seeking protection for a design, but for a unique device, the classifications preceded by a "D" may be disregarded even if the words describing the design class seem to be appropriate.

The second column of numbers may contain the plus sign (+), asterisk (*), or decimal numbers. The plus sign and the asterisk will be explained in Step 3 as the searcher learns about the organization of the *Manual of Classification*, but an explanation of the decimals in the subclass column is useful here. Their origin gives some insight into the history of the PTO and the problems of organizing more than five million documents.

The History of Classification

The patent classification system, developed in the nineteenth century, originally was structured by having class 1, 2, 3, and so on. That part of the classification remains unchanged today with more than four hundred class numbers in existence. As new major classifications were needed, they were simply added at the end. However, subclass numbers within each classification are arranged by technology; for example, machinery that operates on electricity would always be subclassed between subclass 10 and 15 in a particular classification. A problem arose as new technologies were developed. Subclasses 10 through 15 were filled with particular types of electric machines, and then a totally new group of electrical inventions came along that necessitated their own classification, but 10 through 15 were already being used. The PTO could not simply add numbers at the end of the subclass because the arrangement was according to technology; that is, electric patents had to fit somewhere between 10 and 15 unless the PTO rearranged the entire patent classification system.

A suffix was added to the subclass number as a solution. New electric technologies would be classified as 10A, for example. That worked fine until the PTO ran out of alphabet. They then added two-letter suffixes such as 10AA. or 10AB. As more and more patents were added, the classification scheme got complicated and it was possible that a subclass might eventually need to be followed by twenty-six letters. The final solution was to add sub-classifications with a decimal, such as 10.1 or 10.22. The decimal solution is used today and each number is a unique subclassification. But a residual problem exists. The PTO never retrospectively changed all of the old 10A, 10AB numbers, so a searcher will find a mixture of systems in locating sub-classification numbers. Each subclass number found, 10.1, 10A, or 10AB, is a unique subclass, just as if it were numbered 1, 2, or 3.

Returning to the Search

Since looking under Mask proved fruitless, the searcher should use the "see" reference and look in the *Index* under Face Guards. This listing from the *Index* is reproduced in Figure 3–2. What is noticed first is that there is no listing named Face Guards. The correct listing is "Face." "Guards" is listed as a sub-heading. Notice also that Guards refers the searcher back to Masks, but Masks has no listing—the correct listing, as the searcher is already aware, is Mask.

This may seem like nit-picking, but this example points out something important in the editorial control throughout the entire range of patent tools that are used in performing a search. The tools contain many spelling errors, capitalization errors, incorrect references, and downright mistakes! The system is designed for legal professionals who come to the search with a certain amount of background knowledge, but amateur searchers are often thrown off the track and frustrated by not finding information where they have been told they can find it. Be careful. Do not assume that what is printed in the patent tools must be correct. Sometimes it isn't.

Note that listed under Guards as a further subcategory is "Animal restrain-ing type." This may look tantalizing since it mentions animals, but do not be misled. A fly mask is not used for restraining an animal. Now the searcher is at a dead end with Face Guards and Masks, but in Step 1 the invention was broken down and described in as many terms as possible—mask, face cover-ing, flies, and so on. The searcher should try the next term in the list, "Flies."

It seems a little silly to be looking under Flies since insects are not considered inventions. The rational person would disregard this, but the

Face		
False	2	206
Hat combined	2	173
Guards (See Masks)	2	9
Animal restraining type	119	96+
Lifter orthopedic	128	76B

FIGURE 3–2.   Listing for Face

searcher should track down every clue. Although no flies is listed, "Fly" is. In Figure 3–3, the *Index* entry under Fly is reproduced.

Be aware—words listed in the *Index* have no specific meaning. That is, a fly is an insect but to the PTO a fly is an insect, something used by fishermen, the zipper on trousers, the opening in a shoe that holds the laces, or any of dozens of things that could be described as a fly. While the classification scheme itself is organized as to type of technology, such as the electric machines mentioned above, the *Index* is not. It is a true keyword listing without regard to definition.

Look down the list of subheadings under Fly. No Fly Mask or Face Guard is listed, but as was warned above, do not assume that a common term for a device is the term the PTO will use. An experienced patent searcher would see a clue in this list right away, but an amateur searcher would meet with another dead end. The clue will be revealed later in this step.

The next term that was listed in Step 1 is horse. Figure 3–4 shows the listing from the *Index* for Horse. Still no Fly Mask or anything related to it appears, but there are clues here. The amateur searcher who knows something about livestock apparatus would discover it. Instead of being specific about a fly mask, what broad category of horse tack would a fly mask fit into? As Figure 3–4 is being scanned, also notice the subclass number for "Harness" is missing. Only the class number is given—class 54. This happens often in the *Index*, purposely, and means that all types of horse harness may be found in class 54, but a specific item must be further defined to locate its subclass. In Figures 3–2 and 3–3, was anything listed in class 54 that the searcher may have overlooked?

In Figure 3–3 something called a Fly Net is classed in 54. Normally, a searcher would visualize a fly net as something a butterfly collector would use to collect specimens, but remember the warning—terms in the *Index* are true keywords and have no specific definitions. A fly net, because it is in class 54, must be some type of harness.

Experienced searchers would have located this right away, since they usually search only a handful of classifications in which they are expert. They would know that class 54 would be the most obvious place to find a fly mask. An amateur searcher who was the inventor of a fly mask would have found it too, since a fly net is another term for a fly mask.

This is not the end of the *Index* search. Each term that was listed in Step 1 must be located in the *Index*, if possible. The search is continued because

Fly		
Brush	416	501
Closer lasting tool	12	113
Fishing	43	42.24+
Design	D22	125+
Holders for fishhooks, flies	43	57.1+
Net	54	81
Paper box making	493	52+

FIGURE 3–3.   Listing for Fly

Horse		
Blankets	54	79
Design	D30	145
Boots	54	82
Design	D30	146+
Harness	54	
Powered motor	185	15+
Composite or multiple	185	3

FIGURE 3–4.   Listing for Horse

patents do not carry just one class and subclass. Inventions are classed in several different areas that relate to individual components and all possible classifications must be located before moving on to the next step. The *Manual of Classification* in the next step and patent definitions in a later step will both give many classifications to search in locating a particular invention. When the searcher is finished with Step 2, he or she should be armed with a handful of patent class and subclass numbers.

STEP THREE: Using the *Manual of Classification*

The first two steps of the patent search were explained using the *Index to the U.S. Patent Classification*. Before beginning Step 3, the patent searcher should have several possible classifications (class and subclass numbers) that relate to a particular invention. Even if a searcher is reasonably sure that a single correct classification has been found, all related classifications and subclassifications should be explored because a patent is placed in several classifications. One of these is identified as the primary classification and the others as subclassifications. There is no strict rule as to which classification is the primary class. Patent examiners often just pick one as the primary class.

The *Manual of Classification* is the beginning point of Step 3. The *Manual* is a two-volume set of three-ring loose-leaf binders. All Patent and Trademark Depository Libraries and most Government Documents Depository Libraries have the *Index* and the *Manual* in their collections. In the *Manual*, each classification is given one or more pages. The subclassification numbers are listed in two columns on each page and include a two- or three-word description of that subclassification.

In Step 2, a fly mask was used to help explain the use of the *Index* and it was determined that a fly mask was most likely to be placed in class 54, a class titled "Harness."

Figure 3–5 shows part of class 54 as it is printed in the *Manual*. Since this is only a partial listing of what appears in the *Manual*, it is a good idea at this point to look in the actual *Manual* for the full listing.

As explained previously, subclasses are arranged by technology and not numerically or alphabetically. In some subclass listings it may seem that the arrangement is numeric, but on closer inspection some class numbers are not

CLASS 54 HARNESS

1	MISCELLANEOUS
71	BREAKING AND TRAINING DEVICES
72	.Leg spreaders
77	OX YOKES
2	TRACK
24	HALTERS
85	.Connectors
6R	BRIDLES
6A	..Combination halter and bridle
7	.Bits
8	..Mouthpieces
9	...Double

FIGURE 3–5.   Partial listing for Harness

listed in order. The subclass sometimes is taken out of order and placed else-where in the subclass listing. If a subclass is not listed in its logical place, for example, subclass 43 does not come between subclass 42 and 44, look through the subclass listing for it. It will be there—unless it has been changed to another classification (a circumstance that will be discussed in Steps 4 and 5).

The Symbols

Symbols are not explained in the *Manual*. They are explained here in Step 3 because their meaning in the *Manual* is important. (These symbols also appear in the *Index* and are explained in the *Index's* preface.) Figure 3–6 shows the *Manual* listing for Horse.

Figure 3–6 is similar to a listing in the *Index*. Horse/Boots/Design is listed as class D30, subclass 146+. Obviously, this is a design patent in class 30, but the plus sign after the subclass number is puzzling. The searcher should also look at Figure 3–5.

In Figure 3–5, notice how some of the subclasses are printed in all upper-case and how some are preceded by dots? Those in uppercase are immediate

Horse		
Blankets	54	79
Design	D30	145
Boots	54	82
Design	D30	146+
Harness	54	
Powered motor	185	901*

FIGURE 3–6.   *Manual* Listing for Horse

subdivisions of the class; for example, subclass 24, Halters, is an immediate subdivision of class 54, Harness. Those listings that are preceded by dots are further subdivisions of those listings that are uppercase: for example, subclass 85, Connectors, is a subdivision of Halters, which is a subdivision of Harness. Those listings with two dots preceding them are subdivisions of those listings with one dot, and so on.

Class D30, subclass 146+ in Figure 3–6, means that a searcher should look not only at D30/146, but at everything that has dots under 146 in the *Manual* until the next uppercase listing is reached.

For example, in Figure 3–5, if the *Index* had told the searcher to search 54/6R+, the searcher would have to search 54/6R, 54/6A, 54/7, 54/8, and 54/9 because they are all preceded by dots. The search would continue until the next uppercase listing was reached. If the searcher skips Step 3, those additional classifications would be missed.

The asterisk (*) in Figure 3–6 is added to the subclass listings more as a benefit to the patent examiner than to the patent searcher, but understanding it will help the searcher understand the structure of the classification system.

The asterisk is an indication of what the PTO calls a "cross-reference art collection." This has nothing to do with paintings or sculpture. Remember, an art collection to the PTO is a collection of patents relating to a particular technology or art.

For example, buttons may be made by molding plastic, by carving wood, by stamping out metal disks, or any of many methods. Because the subclasses are arranged by technology, each method of button making would be placed in a subclass that relates to the technology used to manufacture it. A patent examiner searching for button patents would have a difficult time in locating all of the appropriate subclasses, so to make it easier the examiner creates an art collection.

Art collections are *digests* set up by patent examiners, and are also official subclasses. Art collections contain only *examples* of button making patents that are placed in the digest by patent examiners to represent the variety of ways in which buttons can be manufactured. When an examiner is considering a button patent, the art collection acts as an index pointing to other classifications where button patents reside. A searcher can use the art collection in the same way to cross-reference an existing patent to several classification areas.

The DIG abbreviation in the Hot Dog listing shown in Figure 3–7 stands for digest. A digest is the same as a cross-reference art collection except that it is *not* an official subclass. It acts as an informal index to other related patents.

This example from the *Index* tells us a lot. It shows a design class, D7, a plus sign indicating that there are subcategories of this subclass that should be searched, and a DIG symbol as a subclass of class 53. The title of class 53,

Hog Dog		
Cooker	D7	323+
Packaging	53	DIG. 1

FIGURE 3–7.   Listing for Hot Dog

which a searcher would learn from consulting the *Manual*, is Packaging. Packaging is also the keyword the *Index* uses to describe the type of patents that belong in class 53. The DIG means that patents in this group, 53/DIG. 1, are patents that relate only to the packaging of hot dogs. They are assigned a DIG category because an examiner has placed a representative sample of hot dog packaging patents here to aid in a search where many different subclasses need to be searched. A DIG subclass may contain only one patent.

In Figure 3–6, the horse-powered motor is an art collection. These are not some kind of mechanical horses, but engines or motors that are driven by horsepower—the real kind that actually uses the horse. Representative patents, whether the horse is running on a treadmill or grinding flour by walking around in a circular motion, would be grouped in this 901 subclassification.

The Rationale of the Manual

A searcher asks at this point why it is necessary to consult the *Manual* at all, since the class/subclass numbers were already given in the *Index*. Is there anything learned in Step 3 that wasn't already learned in Step 2? Yes.

In Step 2 the searcher progressed to the point where there was a good indication that what the PTO called a fly net was the fly mask the searcher was seeking. But there was no verification of that. If the *Manual* showed that a fly net was listed under insect catching devices rather than with harness, the searcher would know that it was a dead end. Without this verification from the *Manual,* the searcher would waste time searching an unrelated group of patents.

Another reason to use the *Manual* is that, in the fly mask example, keywords are listed in the *Index* without regard to definition, so it is often unclear as to what the listing means. The *Index* listing for Fly is shown in Figure 3–8.

What is a closer lasting tool? Could it be some esoteric term for a fly mask? To find out a searcher would consult class 12/113 in the *Manual* and see that a closer lasting tool is grouped with other patents relating to the manufacture of shoes. It is not related to the fly mask that is being sought, but by using *only* the *Index* this information is not clear. The *Index* and the *Manual* should be used together in a patent search. In the Public Search Room of the PTO in Washington, D.C., the *Index* and *Manual* are bound together in special stands throughout the room because searchers use them in tandem.

Figure 3–9 is a reproduction of the listing in the *Manual* where the fly net is listed.

Although the *Manual* is designed to show technological relationships and to assist in the definition of the vague terms listed in the *Index*, this example

Fly		
Brush	416	501
Closer lasting tool	12	113
Fishing	43	57.1+

FIGURE 3–8.   *Index* Listing for Fly

```
                    CLASS 54 HARNESS
         78      TAIL HOLDERS
         79      BLANKETS
         80      BONNETS AND SHIELDS
         81      FLY NETS
         82      HORSE BOOTS
         83R     SPURS
         83A     ..Adjustable
```

FIGURE 3–9. *Manual* Partial listing for Harness

shows that is not always the case. Here fly nets are not defined and they are not a subcategory of another subclassification. The searcher cannot get a handle on what a fly net is, in this instance, other than knowing that it is some kind of harness. The searcher would finish with the *Manual* in the same place they were in when they finished the *Index*—with a handful of class/subclass numbers. But Step 4 shows how to define exactly what a subclass includes.

STEP FOUR: Defining Classifications

In Step 3, patent searchers used the *Manual of Classification* to verify that the class and subclass numbers, identified through the *Index*, were accurate. However, for the fly mask invention used throughout this book, there was no clear indication that what the PTO listed as a fly net was actually the fly mask being searched. The only indication that the searcher was on the right track was that the fly net was grouped with other inventions in a class called Harness.

Step 4 will guide the searcher through *Classification Definitions*, which serve as a dictionary of the class and subclass groups used by the PTO. This step is not always necessary. As mentioned in Step 3, usually the *Manual* gives sufficient information to verify that the classifications about to be searched are accurate, but since the fly mask classification is still vague after completing Step 3, this next step must be taken.

At the PTO, patent definitions are kept on printed sheets. The sheets enable patent examiners to make changes and corrections to the classification by making notations directly on the paper sheets—more about that later. In Patent and Trademark Depository Libraries, the *Definitions* are kept on microfiche cards that are updated periodically, so immediate changes in definitions are not known.

Each microfiche card contains one class. Given that there are about four hundred classes, there are about four hundred pieces of fiche. Each fiche lists the definition for that class and the definition for each subclass within that class.

"Definition" means that the class or subclass is defined in much the same way as a dictionary would define a term, except here the function is explained and examples are given. "Function" is basic to all patent searches and can be best explained by the following example.

Ovens can be of several types: gas, electric, or microwave. Logically, patents related to ovens of any type would be grouped in a class called "Food and Beverage Apparatus." But because of the concept of function, each type of oven is classified with those patents that operate in a similar way. Gas ovens will be classed with other inventions that operate on gas, electric ovens with electric devices that function by resistance in a metal coil causing electron friction to generate heat, and microwave ovens with radar devices that scan with microwaves. Except, of course, where some ovens are grouped in a cross-reference art collection (see Step 3). This explanation is not given here to muddy the waters, but to reemphasize that logically placing similar devices into a single category without regard to function will lead to frustration in your search.

For example, under class 46, "Toys," is the subclass 74. In the *Manual*, subclass 74 is simply explained as "AERIAL," meaning all types of aerial toys are classed here. Refer to Figure 3–10 as the parts of the definition are explained. The *Definition* gives this explanation, "Toys, under the class definition which fly, as ornithopters (flapping wind devices), non-inflatable balloons and dirigibles." This is clearer than simply saying aerial toys in that it gives examples of the type of devices falling under this subclass. Notice that flapping wind devices is a misprint. It should read "flapping wing devices." Searchers were warned of this poor editorial control in Step 1.

Read the definition carefully. It includes the information, "SEARCH THIS CLASS, SUBCLASS: 87+, for inflatable toys . . . ." This means searchers looking for an inflatable aerial toy should check subclass 87+ of class 46 because although inflatables are aerial toys, their operation is different. (Review Step 3 for the meaning of the plus sign.) If the invention was a toy bird that flapped its wings, subclass 134 would be checked. (This is not the same as flapping wing devices in subclass 74. Subclass 134 is just for birds and insects and not for replicas of *machines* that flap their wings.) The searcher is also advised to consult an entirely different class, class 434, if this toy was an educational toy that taught navigation when used with actual pilot training apparatus.

The *Definitions* become a mixed blessing. While pinning down exactly what fits into a subclass, they also give the searcher other places in the classification

---

74. Toys, under the class definition which fly, as ornithopters (flapping wind devices), non-inflatable balloons and dirigibles.

SEARCH THIS CLASS, SUBCLASS:
87+, for inflatable toys, both aerials and non-aerial, subclass 89, having inflatable aircraft simulations.
134, for birds and insects having flapping wings.

SEARCH CLASS:
434, Education devices for training in the operation of aircraft when combined with navigational teaching devices, and subclass.

FIGURE 3–10.   Definition of Toys, subclass 74

to search. This means that if the searcher had only one classification for an invention before checking the definitions, he or she will probably have several others to check when finished.

The Patent Classification System as a Living Organism

A bit of background is helpful at this point. A searcher should not think of the patent classification system as a static list like the Library of Congress cataloging system. The patent classification system is a growing, changing thing. As technologies change or are absorbed by other technologies, the classifications reflect those changes.

In biological classification, a newly discovered animal should be classed with other known animals having similar characteristics and, as much as it can, the patent classification does this. But since completely new inventions occur, new classifications must be added and inventions that previously fit into an existing class might require new classifications that more closely define their functions.

The PTO notifies searchers of these changes in two ways. First, when a change is made in a class or subclass, it is noted at the PTO. On a regular basis, the PTO sends to Patent and Trademark Depository Libraries a listing of what changes have been made—that is, where the class has moved to or what classes have been added or deleted. This is not a rare occurrence; it happens hundreds of times in a given year. This points out that the classification system, like a living organism, is growing and changing daily.

At the PTO, the changes are recorded on 3 by 5 index cards and filed in a metal cabinet near the entrance to the Public Search Room. A searcher who uses the PTO instead of a depository should be aware that not all patents are kept in the Public Search Room. Patent classes do age and die in their own way; entire classifications that show little activity over a given time are removed from the stacks and placed in storage in another building. New subclassifications are continually added, so there are births too.

It is sometimes difficult to envision how such an extensive patent classification system, with more than 400 classes and 104,000 subclasses, does not already have a class designated for any new invention. But consider the photocopy machine. A dry copying process was the idea of a single inventor, Chester Carlson, and was a completely new technology when he got the inspiration in the 1930s. It differed from other copying processes in that previous processes used wet solutions to copy from a master. Carlson's "electrophotography," as the process was called, did not use wet solutions.

Electrophotography reproduced images using an electrical charge to attract the ink to the paper and then heated a plastic "ink" to attach the image to a piece of paper. This was not photography, not wet reproduction, and not anything that fit into existing electronic classifications. A change in the classification system had to be made for this completely new process and in 1942 the first electrophotography patent was granted to Carlson. As mentioned in chapter 1, the patent was granted on the process only, and it was twenty years before the first photocopy machine was built by the Haloid Paper Company (which later became Xerox), but that's another story.

More on Definitions

If the searcher is able to look at the microfiche that contains the definition for class 46, subclass 74, it will help in explaining the next part. Unfortunately, the handwritten changes in the *Definitions* can't be reproduced here.

As automated as modern information is, it is interesting to note that patent classification changes are made manually. On the paper copy of the *Definitions* kept at the PTO, whenever a change is made, one of two things happens: either the affected part of the paper copy is marked with a capital letter in a circle, or a handwritten note is made.

A letter in a circle means that the searcher must locate on the same fiche, usually at the end of the fiche, the matching letter. This matching letter will be the new location of the classification, or a note as to where the classification has been moved. In the class 46/74, the searcher will find a handwritten note correcting the typed text and designating changed classes or changes in wording. When the paper copy is filmed for distribution to the depositories, the handwritten changes and rubber stamps are shown on the fiche.

In the *Definition* for fly mask, class 54/81, the searcher will run into good luck. The *Definition* is direct with no handwritten changes: "Nets for protecting the animal from insects." There are no "Search This Class, Subclass:" notes or "Search Class:" suggestions. This gives the searcher the information needed to continue. Inventions within 54/81 are nets used to protect animals from insects and this is exactly the function of the invention being sought.

The scribbling on the *Definitions* fiche can be disconcerting and confusing to an inexperienced searcher, but using the *Definitions* is worth the effort and will help clear things up when the *Manual* has left the searcher with only a hint of what an invention's classification may be.

Fortunately, when searchers have completed these first four steps they have left behind the most difficult part of the search. Steps 1 to 4 constitute what is called the "Field of Search"—the location of the proper classification.

The next step, Step 5, is the beginning of the "Prior Art" search. Chapter 4 will examine Steps 5 to 7. Prior Art is less confusing, but more time-consuming. Also fortunate, the PTO, with the advent of its CASSIS database, allows a searcher to generate lists of patents within a particular class/subclass group instantly. But as was shown in the Field of Search, the searcher needs to understand symbols and follow procedures.

Chapter Four

Search Procedure Continued: Discovering the Prior Art

The previous chapter guided the patent searcher through Steps 1 to 4, which make up the Field of Search. The Field of Search identifies the proper classifications in which to locate a particular patent. The Field of Search, even for an experienced searcher, is a complex process and carries with it the potential for mistakes and misinformation. Step 5 begins the Prior Art search. Prior Art identifies patents in the designated classifications and leads to actual viewing of patent documents. Prior Art is easier in some ways than Field of Search, but it is more time-consuming.

STEP FIVE: Reviewing the Subclass List

There are two ways to begin Step 5. One is to use the subclass list; the other is to use the Classification and Search Support Information System (CASSIS). We'll look at the subclass list method first.

The Subclass List

Officially, the subclass list is called the Classification Sequence—Subclass Listing or the U.S. Patent Classification—Subclass Listing and may be found in library catalogs under one of these headings. For the sake of brevity, here it will be called the subclass list.

At Government Documents Depositories, the subclass list is on ten reels of microfilm, updated every six months. On each reel the class and subclass sets are given in numerical order, and the patent numbers that are classed in that set appear under each. For example, the subclass list for 46/74D would look like Figure 4–1 below.

In the subclass list, design patents are placed first. As mentioned before, design patents are designated by a D before the patent number. An explanation of design patents was given in Step 2. After the design patents, other patents are arranged in numerical order—the lower the number, the older the

```
┌─────────────────────────────────────────────────────────────────┐
│  PATENT CLASSIFICATION - SUBCLASS LISTING          REEL NO. 2    │
│  ───────────────────────────────────────────────────────────    │
│     74D                                                          │
│   D 162,560U                                                     │
│   D 183,626U                                                     │
│   D 214,577X                                                     │
│        .                                                         │
│        .                                                         │
│        .                                                         │
│     3,113,396                                                    │
│     3,359,678                                                    │
│     3,545,760X                                                   │
│        .                                                         │
│        .                                                         │
└─────────────────────────────────────────────────────────────────┘
```

FIGURE 4–1.   Subclass list for 74D

patent. After the patent number is a U, an X, or nothing. The X means that this patent is a cross-reference patent.

In previous steps the searcher was told that patents are not placed in just one classification, but in several. One classification is the primary class and the others are secondary or cross-reference classifications. A patent number with an X means that this is a secondary classification. If the searcher is searching more than one class/subclass, this patent number will appear again.

A U is an "unofficial" secondary subclassification. The usage of any secondary classification is similar to the DIG category mentioned in chapter 3. An examiner places this patent in this subclassification to act as an index to the many ways some things can be classified. Because of this a searcher may find patents marked with a U that, when viewed, seem to be only marginally related to the other patents in the subclass. For example, the class/subclass list containing the fly mask, 54/81, also contains a patent for a potato digging machine. The potato machine patent is there because something about it relates directly to the class Harness; perhaps the fact that it is attached to a horse by means of a harness. The patent examiner wanted to be sure to cover this aspect when harness patents were being searched.

The U designation is no longer used. Previously, examiners were allowed five official cross-references to be added to the original classification. If an examiner felt that additional cross-references were needed, the U designation was used. In current practice, any classification in addition to the original classification carries an X designation. The U classifications are being phased out, but they still will appear on subclass lists since the PTO is not removing them retroactively. A searcher should treat X and U as equals of the primary classification and search all numbers listed on the subclass list.

Keeping Records and Attorney Fees

The searcher needs to keep accurate records of the patent numbers that have already been viewed so that when the X number found in the cross-reference appears again in the listing of primary class patents, it is not necessary to view it again.

There is another reason to keep accurate records of what has been searched. Many searchers will complete the Field of Search, the subclass list, and a partial viewing of patents on their own, but hire an attorney to interpret certain complicated patents or to complete the search of patents in a subclass list. The attorney will want a record of what has been done up to that point. If the record is not complete and accurate, the attorney will duplicate the search already done and charge an hourly rate, which defeats the savings of do-it-yourself searching.

Attorney's services and fees vary greatly according to geography, the type of patent being searched (some take longer than others), whether the attorney is handling the entire procedure or just a part of it, and the amount of haggling the attorney and the patent examiner get into when the application is examined. In all cases, however, the attorney will have to be a registered patent attorney. Lawyers are required to pass an examination given by the PTO before they may practice patent law. A listing of these patent attorneys can be found in most Government Documents Depository Libraries and at all Patent and Trademark Depository Libraries in a volume titled *Attorneys and Agents Registered to Practice before the United States Patent and Trademark Office.* A somewhat easier way to locate a patent attorney is to look in the yellow pages under "Patent Attorneys."

Patent agents may also perform patent searches and file patent applications. A patent agent must pass the same written test as a patent attorney and also must have a specified amount of professional training in engineering. The difference is that a patent agent is not a lawyer. An agent is trained in patent law and procedures, but has not passed the bar exam.

Patent examiners, those employees of the Patent and Trademark Office (PTO) who determine whether an invention is new, need not be attorneys either. Patent examiners are among the best-educated people in the federal bureaucracy, with many holding doctorates in technical fields. As mentioned earlier, a problem in recent years is that patent examiners frequently attend law school in the evenings at government expense, then leave the government and practice patent law.

The patent attorney fees shown in Figure 4–2 are representational and may vary depending on the variables mentioned above.

Service	Time	Cost
Patentability search and Prior Art search	4–5 weeks	$200-300
Patent application	3–4 weeks	$1,100
Prepare drawings		$80/drawing
Filing fee		$170
Office action prosecution	10–12 months	$400+
Final issue fee	18–24 months	$340

FIGURE 4–2.   Representative patent attorney fees

The office action and the filing fee run concurrently; that is, the filing fee is paid about twenty-four months after the patent application is filed, not twenty-four months after the office action. Using these figures, even if a search was done completely without an attorney, an inventor would still have to pay $2,330 (allowing for four drawings at $80 each) if a patent attorney was used to prepare and file the patent application.

CASSIS

A subclass list also may be generated through a system called CASSIS, the Classification and Search Support Information System. CASSIS is a compact disc index that runs on a personal computer and is available at the PTO and at all Patent and Trademark Depository Libraries (PTDLs). It is the quickest way to obtain a hard copy of an up-to-date subclass list. The searcher simply types in the patent class/subclass. The list of patent numbers will look essentially like the film list in Figure 4–1.

Although CASSIS is fairly simple to use, searchers who have never tried it before will need an introduction. Librarians at PTDLs, the only libraries with CASSIS access, are trained in its use, so a discussion of what CASSIS does and how it does it will not be detailed here. Librarians not trained in CASSIS should consult a nearby PTDL (see Appendix 2) for instruction.

CASSIS does many things well for patent searchers. If a searcher has a patent number, CASSIS can generate a list of all of its classifications. This is helpful if a searcher knows of an existing patent that is similar to the one being searched. After a searcher inputs that patent number, CASSIS prints out the possible class/subclass groupings. This is a real time saver from the usual field of search procedure.

CASSIS also can display a page from the *Manual,* display only those subclasses having the same level of indentation (see Step 3 and the dots preceding a subclass), search keywords within classification titles and in patent abstracts, search terms that appear in the *Index* (see Step 2), and, in more recent patents, display patent titles and names of companies having rights to the patent.

What to Do with a Subclass List

When a searcher obtains a subclass list of patent numbers, the next step is to obtain those patents on the list. Given that a searcher has one to ten subclass lists and each list contains at least one hundred patents, this is a formidable task. Described below are the four methods of wading through this list.

The OG Method

The *Official Gazette* (OG) for patents is published each Tuesday. In it the patents granted during the previous week are listed in numerical order and grouped by class/subclass. Although the entire patent is not reproduced in the OG, it contains a brief abstract and usually a reduced patent drawing. This is a popular search technique because more than 450 libraries receive the OG and the patents may be scanned quickly. The spine of the OG lists the patent numbers included in that issue.

The CASSIS Method

Patent titles for patents issued between 1969 and the present are viewed on a printout. Titles that seem unrelated to the one being searched are eliminated at this stage before doing the actual viewing of the patent.

The PTDL Method

Actual patents, complete with drawings, are reproduced on rolls of microfilm. PTDLs buy the film from a commercial vendor and are required to have at least twenty years of patent backfiles. However, dealing with film and the numerical arrangement of the patents, as opposed to a classification arrangement, is frustrating and time-consuming.

The PTO Method

This is *the* preferred method, although its obvious disadvantage is that one must travel to the PTO in Washington, D.C. Patents are reproduced in paper booklets that are printed at the time the patent is issued and are grouped at the PTO into class/subclass arrangement. All patent attorneys and agents use the PTO method.

STEP SIX: Locating Patents by Number

In Step 5 searchers were shown how to obtain a patent subclass listing. The subclass listing identifies patent *numbers* appearing in a given class/subclass group. Since only the number of the patent is given, the next step shows how a searcher may locate the actual patent by using the patent number.

Obtaining the Patent

Step 6 is actually a set of alternatives to secure the patent document. The primary consideration in choosing a search method is the type of patent depository the searcher is using. There are three types of patent depositories. The most common is a Government Documents Depository Library. The others are Patent and Trademark Depository Libraries (PTDLs) and the Patent and Trademark Office (PTO) itself.

Patent Depositories: Three Types

Government Documents Depositories

These are located in more than 450 libraries in the United States. The depository collects documents published by the federal government and makes them available to the public. Each library collects a percentage of those things that the government identifies as depository items. Some standard depository items are the patent tools that have been described previously, e.g., the *Index*, the *Manual*, and the *Official Gazette.* These libraries will not carry actual patents, but with the OG a patent search may be initiated. The OG method below shows how to use the OG to locate patents by number.

Patent and Trademark Depository Libraries (PTDLs)

More than sixty PTDLs are scattered across the United States. They differ from Government Documents Depository Libraries in that the PTDLs carry a backfile of at least twenty years of patents. The patents are on microfilm rather than in paper copy and are arranged by patent number instead of by classification. The PTDL search method described below is the most convenient for patent searchers, but is very time-consuming.

The Patent and Trademark Office (PTO)

The PTO is the only practical place to perform a comprehensive patent search. However, the PTO is not convenient for the majority of patent searchers since it is located in Washington, D.C. Most searchers find that a search initiated at another location and completed at the PTO is most effective.

Advantages and Disadvantages of the Four Search Methods

Four sources facilitate the location of patents. Each source with its accompanying search method (described below) is used by itself or with other methods to locate a patent by its number.

The CASSIS Method

CASSIS—available only at PTDLs and at the PTO—was described in Step 5 as an automated system that provides, among other things, a listing of patent numbers in a given class/subclass group. In addition to giving patent numbers, in listings of recent patents, CASSIS also gives the title of the patent. This feature can be used by a searcher for a quick elimination of patents that, by their title, have no relation to the invention being searched. All unrelated patents cannot be eliminated this way, but this method is effective in reducing the number of actual patents that need to be viewed in their full format at a PTDL or at the PTO.

Of course, this can be risky because patent titles sometimes give no indication of what the patent is. Also, the CASSIS method is not a technique for a comprehensive search since CASSIS only lists titles for about the last thirty years in most classifications. But it is a useful tool for reducing hundreds of patents in the group of class/subclass areas that a searcher is examining. And any reduction in the number of patents to be viewed can save many hours of search time.

The OG Method

The OG method uses the *Official Gazette of the United States Patent and Trademark Office*. The advantage of this method is that it can be used at any patent depository location. As mentioned before, the OG is published every

Tuesday and contains all of the patents issued during the previous week. However, it does not contain the actual patent but rather an abstract of what is patented and usually a single drawing of the invention.

Armed with a list of patent numbers, a searcher would go through each weekly issue of the OG and eliminate unrelated patents in much the same way as the CASSIS method. The advantage here is that the searcher is working not only with a title, but also with a description and a picture. The OG method is also not a comprehensive search method, but like the CASSIS method its purpose is to reduce the number of actual patents that need to be viewed in their full format.

The obvious disadvantage is the wealth of material. With fifty-two OGs published each year, even a minor search of patents in a given class/subclass group for the previous ten years would mean that a searcher would be handling 520 OGs—a very time-consuming and tedious task.

The PTDL Method

The CASSIS and OG methods can be used at a PTDL, but another method is unique to PTDLs. Once a searcher has a list of patent numbers, the PTDL allows the searcher to go directly to the patents for viewing. Patents at PTDLs are on microfilm and not in a paper format (with only two or three exceptions). The microfilm is arranged by patent number, so the searcher need only locate the reel of film that contains the number being sought. The film is a full duplication of the paper copy of the patent and contains all drawings. As mentioned above, this is the most convenient method for most amateur searchers because most states have a PTDL.

However, the PTDL method has two major problems. The first is that PTDLs are required to keep a backfile of only twenty years worth of patents. If a patent is good for seventeen years, this seems to be a sufficient backfile, but as professional searchers know, it's not. If a device has been patented previously, even though the patent has expired, it cannot be patented again, so a searcher must search back as far as the technology might have existed. If a searcher is looking for electronic equipment patents, twenty years may be sufficient since the technology is relatively new. But in the fly mask example that was searched in the first several steps of this series, a searcher would have to go back to 1790, when U.S. patenting began, since fly masks have existed for centuries. Some PTDLs have complete patent files, but others have patents only from the 1960s.

The other major problem is the film itself. PTDLs chose this method because it was easy to update patent files by simply adding another reel of film with the most recent patent numbers on it. If the film was arranged in class/subclass order, the entire set of patent film would have to be replaced each Tuesday since new patents would be added throughout the classification system. With the PTDL film system, a searcher looking for one hundred patent numbers might have to handle one hundred different reels of film.

An experienced patent searcher needs to locate the film reel, load it on the film reading machine, roll the film to the appropriate location, view the patent, and rewind the film for each patent. The rate at which this can be done is about seven patents per hour. At this rate it would take more than fourteen hours to view one hundred patents. In a normal patent search, a searcher easily views three hundred or more patents.

The PTO Method

This is the only comprehensive, simple method of searching a large number of patents. At the PTO, paper copies of patents are arranged in class/subclass groups and placed in their own compartments called "shoes." A searcher has only to walk to open shelving containing the shoes, pull the appropriate classification, and easily flip through the paper copies to perform a search. The PTO even provides custom-made V-shaped stands to allow the patents to be viewed vertically.

A searcher should be aware that although the common belief is that all patents are housed in the Public Search Room of the PTO, they are not. The fly mask classification 54/81, for example, is considered an inactive classification and is housed in another building in Crystal City, but it is in paper and readily available.

Weighing the Prior Art Search

The Prior Art search is important, but even an amateur searcher should weigh its costs. A searcher makes a search to verify that an invention is not already patented. There is no penalty for submitting an application to the Patent Office for a device that is already patented; however, the money spent for making the application, preparing the drawings, and the time spent in preparing the application will be lost if the PTO finds that a patent already exists for the invention being submitted. An inventor who has performed the search and prepared the application materials without assistance from a patent attorney or agent would lose about $600.

Some inventors are willing to risk an application without performing a search at all, although the PTO requires that a listing of class/subclass groups searched be included on the patent application. This is not as odd as it sounds. Even though an amateur searcher or an attorney has performed a comprehensive search, the PTO performs its own search. If the PTO finds an existing patent, the applicant will lose all fees. An inventor must weigh the potential risk of losing the application fees against the time or money spent in searching prior patents.

Eliminating Unrelated Patents

After deciding on a search method, a searcher should eliminate patents that are not related to the invention being searched. This part of the search requires some skill and familiarity with the technology that created the invention being searched. Being able to view a patent and make a decision quickly on its relationship to another invention is a valuable skill.

Take the example of the fly mask invention. The searcher should ask, "What is it about this particular fly mask that makes it unique and patentable?" Is it made with a unique design? Is it fastened to the animal in a unique way? Does it adjust to fit all sizes of animals?

If an invention has several unique features, the searcher should focus on one item at a time until getting the hang of searching. The fly mask being searched has one unique thing about it—the straps that hold it in place are designed so that the mask can fit any type of animal whether a sheep, cow, or horse. This then is the focus of the search.

Amateur searchers have one advantage over professional patent searchers. The amateur searcher is likely to be the inventor of the device. The unique features of an invention are well known to the inventor, and the inventor is usually working with one particular technology with which he or she is very familiar.

For example, the inventor of the fly mask knows how a strap must be fixed to the mask and which features make a strap adjustable. When the inventor-searcher is going through class 54/81, the fly mask classification, it is readily apparent to him or her whether an existing patent has the unique features of the new invention.

Patent attorneys usually do not search the patents at the PTO themselves. They hire professional searchers in the Washington, D.C., area to perform searches for them. These professionals search a wide spectrum of classifications. Although most professional searchers are very good at what they do, they didn't invent the fly mask. They must depend on documentation from and communication with the patent attorney to explain what is unique about the patent being searched. The amateur is usually quicker and more aware of subtleties in existing patents.

Assume that the amateur has followed Steps 1 to 6 and has identified a class, has a list of patent numbers, has eliminated a number of patents that seem unrelated by using the OG, and is now seated in the PTO Public Search Room. Just what is it in the patent that a searcher is looking for? Specifically, two primary items: the drawings and the claims.

STEP SEVEN: Examining Patent Claims

Claims are the effective part of a patent. They are numbered paragraphs that give a precise description of the invention and list all essential features. The claims are the basis for a patent infringement suit in that what is unique about an invention must be mentioned in the claim.

In disclosing a patent (i.e., publishing the patent complete with drawings, descriptions, claims, etc., as is done when a patent is granted), occasionally a unique feature is not claimed although it appears in the drawings. David Pressman, a noted patent writer, states that one may copy that part of the invention without liability since it is not described in the claims. Obviously, it is important that every unique feature of an invention is accurately described in the claims.

Claims define the structure of an invention in precise terms. The legal protection given the patent is delineated by the claims and not the drawings, disclosure, specifications, or any other part of the patent. Patent attorneys and agents are trained in the art of writing a claim narrow enough to show uniqueness, but broad enough to give an inventor some protection over other similar inventions. The importance of the claims cannot be overstated. So what does this mean to a patent searcher, and how can it help in the search process? In our example, the unique features of the fly mask can be found in the claims.

In the last step it was determined that one of the unique things about the fly mask invention being searched was that it had straps that enabled it to fit any type of animal. In the viewing of the actual patent, a searcher could look at the drawings to see if an existing fly mask patent had straps and, if it did, read the claim to see if the straps were adjustable and, if adjustable, determine if they adjusted in the same manner as the fly mask being searched. The variations can be extensive. For example, are the adjustable straps connected with Velcro or with a buckle and does that make a difference in infringement? Inventors should make that determination for themselves. They cannot depend on a librarian or other facilitator to make it for them.

Note to librarians: Be careful in explaining patent claims. Interpretation of claims gets into the fuzzy area of patent law, and a decision as to whether a particular claim is an infringement on an existing patent is a matter best left to patent attorneys and agents. Interpreting claims in other than a very broad explanatory sense constitutes the practice of law. Librarians and others assisting with a patent search must be careful to explain what the claims are, not interpret them. Medical librarians are familiar with this type of precaution, since they cannot diagnose an illness for a library user or recommend treatment.

One of the claims in patent 3,803,801 follows. Since patents are public documents, they are not copyrighted. It requires no permission to quote directly from them. This patent for a fly mask, granted in 1974, is one of only five patents in class 54/81 granted in the past nineteen years.

What I claim and desire to secure by Letters Patent is:
1. An insect control device for animals and comprising:
    a. a band adapted to be placed around a neck portion of an animal;
    b. a nose strap secured to said band and adapted to extend at least partially around a nose portion of the head of the animal;
    c. a head strap connected to said band and adapted to extend at least partially around the head of the animal and adjacent to the eyes thereof;
    d. a plurality of laterally spaced narrow substantially rigid strips mounted at one end thereof on said head straps and positioned to extend at least partially over the eyes of the animal;
    e. said band and nose strap and head strap are each adjustable to conform to the respective portions of the head of the animal; . . .

In 1.e., the patentee states that this fly mask is adjustable. The searcher who sees this should immediately take a look at the drawings that accompany the patent. Patent 3,803,801 has five drawings. One of them clearly shows an adjustable strap that connects by means of a buckle like those on belts. Is this adjustable strap close enough to the patent being searched that it constitutes an infringement? No decision can be made by a librarian. The searcher is also wise not to make a definite judgment at this point, but to place the patent in a group of questionable patents that will be generated during the search. This group of questionable patents should then be taken to a patent attorney or agent for an interpretation.

Patent 4,791,777 is the last patent granted in class 54/81. It was awarded on December 20, 1988, and has only a single claim:

1. A leg net assembly positionable about the leg portion of an animal to prevent insects from contacting the leg, said leg net assembly comprising:
    a central, generally cylindrical, loose fitting body portion . . . and upper and lower supports, said upper and lower supports each being formed of a knit, high stretch material and being attached to upper and lower ends, respectively of said central generally cylindrical body . . .

It is obvious that this fly net is designed to cover the legs and not the face of an animal. It is also obvious that the attachment to the animal is by stretch fabric bands and not buckles or straps. A searcher will probably disregard this patent immediately as being unrelated to the fly mask being searched. It is important for the librarian who may be helping a patent searcher to remember that no interpretation can be made; that is, if asked by the searcher if this patent is related or unrelated to the fly mask being searched, the librarian can give no answer. It is up to the searcher or the searcher's attorney to make that decision.

At this point in the search process, the librarian should let searchers proceed on their own. Interpretations, writing the patent, filing forms, and making drawings are things that are best left to inventors and their legal representatives.

An amateur patent searcher will leave the library having completed the search process and will have a small group of patents that can be shown to an attorney for interpretation.

Summary of the Process

The major steps to searching a patent follow:

Step 1: Identify the parts of the invention.
Break the invention down. For example, a generator-driven electric light for a bicycle would include a light, a generator, and a bicycle. Think of as many synonyms or related terms as possible to identify the parts of the invention.

Step 2: Consult the *Index to the U.S. Patent Classification.*
The keyword *Index* will give a class and subclass for each term identified in Step 1. When done with this step, a searcher will have many sets of class/subclass numbers.

Step 3: Consult the *Manual of Classification.*
The *Manual* will show the relation of the invention to other inventions within the same technology. It also provides the searcher with finer indexing of specific aspects of the invention leading to more direct class/subclass listings.

Step 4: Consult the *Patent Definitions.*
The wording in the *Index* and *Manual* is sometimes vague. This leads to some confusion as to whether the classifications the searcher has identified are

correct. The *Definitions* specifically state what type of device fits into every class and subclass. This is sometimes as confusing as the *Index* and *Manual* themselves, so it's good to know that many times this step can be skipped if searchers are fairly certain they have the right classes.

Step 5: Get a list of patents in each class/subclass group.
Once the classifications are identified, the list of patents in each class can be generated through CASSIS or through the subclass list on microfilm.

Step 6: Eliminate each obviously unrelated patent.
Searchers can do this four ways: by going to the PTO in Washington and viewing actual patents, by scanning the OG, by scanning a CASSIS listing of titles to eliminate obviously unrelated inventions, or by using a PTDL to view patents on film.

Step 7: Consult the claims to eliminate unrelated patents.
Claims state specifically what is new about the invention. By reading the claim, most patents can be kept or disregarded as unrelated to the invention being searched. Those patents that are of questionable relationship to the patent being searched should be collected and a decision made as to their relevance by searchers themselves or by patent attorneys or agents.

One comfort in all of this is the knowledge that the PTO will perform its own search to verify that a patent does not already exist. The searcher goes through this process because if the PTO finds a patent that the searcher has overlooked, the inventor will forfeit all application, preparation, and legal fees. As said before, the search should not cost as much in fees and the searcher's time as the inventor stands to lose if an existing patent is found by the PTO.

It Can Be Done

Step 7 completes the librarian's role in offering assistance to an amateur patent searcher. By following the steps outlined in chapter 3 and here in chapter 4, and by remembering the background knowledge presented, a librarian should be able to make a searcher feel confident and sure of the basic procedure in applying for a U.S. patent. Of course, the librarian can point to books, articles, and supplemental materials that can assist a searcher. For selecting an attorney, the searcher can be shown the list of registered patent attorneys and agents, for example.

Many patent writers say that the U.S. patent system has effectively eliminated the individual inventor by making the process complicated and expensive. A patent search can be an intimidating and confusing process to a novice. But while the system is not designed for amateurs and they are at a disadvantage using it, thousands of amateur searchers each year successfully perform their own searches. Unless the invention being searched is extraordinarily complicated, such as some recent electronics devices, even those with modest educational and legal backgrounds can succeed in completing the process.

The U.S. Constitution wisely guarantees the right to patent for individuals and not for corporations. Unfortunately, in the twentieth century, unlike previous times when a patent on a simple idea could make an individual wealthy, it is rare today to find individuals who are able to profit from patenting a simple device. However, the government through its PTDL program, which makes patents available to the public, and private organizations through programs such as Invent America, which encourages school-aged children to invent patentable devices, are making an attempt to preserve the spirit of American ingenuity.

The U.S. patent system is a living, changing thing. The process outlined in this series may change in the next few years as the PTO attempts to apply new technologies, such as compact disc storage to the system. Nor is the patent system free of controversy that may change search techniques or the manner in which patents are secured. Librarians should try to be aware of these changes as they occur.

Chapter Five

Patents, Copyrights, Trademarks, and Trade Secrets

Librarians, the public, and even legal professionals are often confused about the similarities and differences between patents, trademarks, and copyrights. It is common to hear laypeople refer to "patenting a book" or "copyrighting an invention." Before exploring the distinctions of patents, copyrights, and trademarks, let's look at a basic description of each.

Copyright grants protection for artistic expression. Most people are familiar with the concept of artistic expression as a written work like a book, but copyright also protects jewelry designs, music recordings, and sculpture, among other things. In this chapter, it is important that the reader think of copyright in this broader sense.

A patent protects a machine, process, manufacture, or composition of matter. Essentially, a patent protects things that have a use or are utilitarian; copyright (in the broader sense above) protects things that have no utilitarian purpose.

Trademarks protect a word, name, symbol, or device that is used in trade to indicate the origin of goods or to distinguish the goods from the goods of others.

A trademark is a word, words, or graphical symbol or design that identifies a person's goods and distinguishes them from others. A trademark may be a coined word, such as Coca-Cola or Xerox, or a word or phrase used exclusively in tandem with a product, such as "You Deserve a Break Today." A rule concerning a trademark's use is that the word or phrase used must not be generic. For example, if the product was an electric fork it could not carry the trademark "Electric Fork." A trademark's purpose is to fix in the mind of the consumer a relationship between the trademark and the quality of the goods or service, or to make goods or services easily recognizable and easily distinguishable from other goods and services of the same type.

Trademarks are different from patents and copyrights primarily in that *a trademark must be used first on a sale of a good before it is eligible for registration.* A further requirement of a federal trademark is that the goods or service must be used in commerce regulated by Congress, that is, sold across state lines. This is because the Constitution does not specifically protect trademarks. They fall under the Commerce Clause of law and so are protected by the federal government. State trademarks also exist. These trademarks offer protection only in the state in which they are granted and are available for a nominal fee from the state's Secretary of State office.

A trade secret is not a registered form of protection. Trade secrets protect a device, process, or formula by simply keeping the item secret. The best example of this is the formula for Coca-Cola, which is said to be held in a vault. If the formula for Coca-Cola were patented, it would have to be revealed and all rights to the formula would have expired decades ago.

Applying the Concepts of Intellectual Property

Most of the confusion between these types of protection arises from the similarities between copyrights and design patents. Design patents are one of the three types of patents (see chapter 1) and are similar to copyrights in that both can protect designs or ornamentation; however, a primary characteristic separates copyrights and design patents. Just as utility patents must be useful or utilitarian, design patents must be part of a useful device. Copyright, on the other hand, is granted to designs or ornamentation that are not utilitarian. For example, a chair of a unique design would be granted a design patent because the design is part of a useful item; a copyright would be granted a drawing of the chair because the drawing itself is considered to perform no useful function. Title 17 USC 101 says that a useful article is an article having an intrinsic utilitarian function that is not merely to portray the appearance of an article or to convey information.

Copyright

Copyright gives the author of a written work the exclusive right to copy or reproduce his or her writings beyond certain "fair use" provisions. Copyright, however, does not apply only to writing. Copyright protects musical works including lyrics, dramatic works including accompanying music, pantomimes, choreography, pictorial works, graphic works, sculpture, motion pictures and other audiovisual works, sound recordings, and some computer programs.

Forms of expression not copyrightable are: works that have not been fixed in a tangible form, such as an improvised speech, song, or performance that has not been written or recorded; names or short slogans such as bumper stickers or sayings on T-shirts such as "Where's the Beef?" (see trademark section below—"Where's the Beef?" is a registered trademark); listings of ingredients; works consisting of information that is common property and contain no original authorship; works of the U.S. government; and works in the public domain.

One of the more important aspects of copyright as opposed to patent protection is that the creation being protected must be nonutilitarian. Items such as blueprints or architectural designs, jewelry, dolls, and designs on fabrics would be copyrighted rather than patented because they perform no function and are not "useful."

Copyright protection is for the *form* of expression rather than the *function* or subject matter. In other words, copyright only protects expressions of ideas, not the ideas themselves. An author can draw substantially from the works of others and still receive copyright protection for a work as long as it varies distinguishably from other works and represents the author's exercise of skill, labor, and judgment. For example, one artist's drawing of a chair could be copyrighted and another artist's drawing of the same chair could have a separate copyright, even though both are drawings of the same thing.

Design Patents

Design patents, one of the three primary types of patents, protect the appearance of an invention and not its function, although the thing being protected by a design patent must be a utilitarian item. Again, this is the primary distinction between patents and copyrights: artistic expressions are copyrighted, utilitarian things are patented.

Think of a chair. In designing a chair that had, say, a single pedestal instead of four legs the inventor is not reinventing the chair. It still performs the same function of supporting a person in a sitting position. The inventor is just changing the way it looks; however, since a chair is a *functional* item it is protected by a patent rather than a copyright. Design patents have been granted on articles such as belt buckles, shoes, typeface designs, and watch faces—all functional items.

An "invention" must meet more stringent requirements to qualify for design patent protection than copyright protection. While copyright only requires the *expression* of an author or artist, a design patent must be original, new, and ornamental, and be *embodied* in an article of manufacture. Design patents must also satisfy the general patent tests of novelty and non-obviousness (see chapter 1) and must rise to a level of invention beyond the basic level of creativity required for copyright protection.

Comparisons

Basically, design patents apply to completely new, original ornamentation that appears on a utilitarian article, while copyrights apply simply to the artist's expression of a nonutilitarian design, whether it is new or not, or to a design that can exist independently of the utilitarian aspects of the object.

Court Cases Involving Patents and Copyrights

An example of how the courts perceive the distinction between design patents and copyrights was shown by the Supreme Court in the case of *Mazer v. Stein*. Stein created molded statuettes and registered the design as a nonutilitarian "work of art" under the copyright law. Mazer copied the statuettes and by drilling holes in the top and bottom and wiring the sculpture, rendered the statuette a functional lamp. Mazer argued that since the work of art was functional in its use as a lamp base, the copyright law would no longer protect

Stein. The Court ruled in favor of Stein and held that even though a functionally useless item could be rendered functional, that action cannot negate protection of the functionless sculpture under copyright law.

King Features v. Kleeman demonstrates the concept of copyright protection as it applies only to the specific thing being protected as in the example above of two artists making a drawing of a chair. King Features copyrighted "Popeye the Sailor" originally as a comic strip, a two-dimensional graphic work of art. Kleeman used the two-dimensional comic strip drawings as a guide in manufacturing three-dimensional Popeye dolls. King Features claimed infringement of copyright and won the decision, but only after they were able to show that the alleged infringements were duplicates of actual copyrighted drawings. Had Kleeman made the dolls from a Popeye drawing that Kleeman had created himself, the decision may have been in Kleeman's favor.

Patents, Copyrights, and Trademarks

The Rolls-Royce automobile hood ornament, generally known as the "Flying Lady," was created in 1910 by a famous sculptor of the time, Charles Sykes. Sykes originally named the sculpture "The Spirit of Ecstasy" after riding in a Silver Ghost Rolls-Royce. The sculpture functioned as the automobile's radiator cap and, since it was a functional design, was registered as a design patent by the Patent and Trademark Office (PTO). It was also protected by a copyright as a work of art in its original creation as a statuette. The Flying Lady is also the registered trademark of Rolls-Royce (see the trademark section below.)

An interesting point concerning items that may be both patented and copyrighted is that the artist must obtain the patent first. The rule described in chapter 1 explains that in order to qualify for patent protection, the invention must not have been shown previously in any publication. If copyright were obtained first, this would make the invention public and negate the artist's opportunity to patent. Some courts have ruled that if protection is granted under copyright, a design patent cannot be later granted. However, if a patent is granted to a design, it may later be copyrighted since copyright law has no provision for the novelty of the idea.

Considerations

Several other important points help in distinguishing between copyright and design patent protection. The first is length of time necessary to secure rights. Copyright registration is granted in a few weeks while design patent protection requires six months or more. The additional time taken in a design patent request is usually for the extensive search of existing patents and the preparation and filing of the many documents required by law. A copyright, however, is not searched and requires only one simple form. The Copyright Office makes no attempt to screen copyrighted matter unless the work of art obviously cannot be protected because it is utilitarian or "scandalous or in some other way objectionable." The Copyright Office will normally accept the fee and register the copyright.

A design patent is granted by the PTO, a division of the U.S. Department of Commerce, while a copyright is granted by the Copyright Office of the

Library of Congress. Separate laws cover design patents and copyright. Copyright is covered in the Copyright Act of 1976 (public law 94-553, October 9, 1976) (25) and Title 17 of the U.S. Code, while design patents are covered, for the most part, in Title 35 of the U.S. Code.

The length of time rights last is also significantly different between the two types of protection. According to the Copyright Act of 1976, copyright lasts for the life of the author plus an additional fifty years. The term of a design patent, initially set at seven years, was changed in 1861 to be three-and-one-half, seven, or fourteen years, chosen at the discretion of the applicant, with fees varying according to term. Design patent provisions, as amended in October 1982, now fix the term at fourteen years.

An important concept in the copyright law concerns the type of rights granted to the author. Copyright protection grants certain exclusive rights to distribute the work of art, to copy it, and to display it publicly. But it does not grant exclusive right to manufacture the work of art nor does it grant protection of the utilitarian aspects of the article. Pictures of the work of art appearing in news reports, for example, would not ordinarily infringe upon the rights of the copyright holder. In comparison, design patents grant the owner the right to exclude others from making, using, or selling the invention. Thus, design patent protection confers to the owner a temporary monopoly over the manufacture, use, and sale of the invention.

Assuming that a design can be judged as a work of art, its protection under the copyright law is quite simple. Two photographs of the work of art can be sent to the Copyright Office along with the appropriate form (Form VA for visual arts and form TX for written works, available from the Copyright Office) in duplicate, with a small fee. The copyright in the form of a stamped application is usually returned to the applicant in several weeks.

A design patent is not as easy to obtain. It is expensive and will cost the patentee several hundred dollars. As explained in previous chapters, a patent cannot easily be secured by a do-it-yourself application. The preparation of the application is a highly technical job, not a simple fill-in-the-blanks application, and is better left to a patent attorney or agent. The drawing that is required with the patent application should be prepared by a professional patent draftsman. The patent is granted in about a year on the average.

Figure 5–1 highlights the major differences between design patent and copyright.

Trademarks and Trade Secrets

A few examples can illustrate the importance of a trademark. A strong visual association exists between certain products and their trademarks. Some of these are the Smith Brothers on cough drops, McDonald's Golden Arches, and Planter's Mr. Peanut. Phrases such as Wendy's "Where's the Beef?" or Coca-Cola's "The Pause that Refreshes" are trademarks, since short phrases cannot be copyrighted.

In some cases, color can be a trademark, as in the case of pink being used exclusively by Owens-Corning for fiberglass insulation; however, pink was not allowed protection by Pepto Bismol since it was thought that it is a soothing color and therefore serves a functional purpose. In a similar case John Deere

	Design patent	Copyright
WHAT IS PROTECTED	Ornamental designs for an article of manufacture	Writings, photos, music, works of art, computer programs, architectural works
CRITERIA FOR PROTECTION	New and non-obvious, serves a functional purpose	Originality, non-utilitarian
PERIOD OF PROTECTION	Fourteen years	Lifetime of the author plus fifty years
APPLICATION PROCEDURE	Complicated forms	One simple form
FEES (at this writing)	$70 plus other costs, possibly hundreds of dollars	$10

FIGURE 5–1.   Major differences between design patent and copyright

was refused trademark on the color green for farm equipment because it would prohibit other equipment manufacturers from using green—a color closely associated with agriculture.

Unlike patents and copyrights, trademark protection can be diminished through excessive public use, which permits the trademark to enter into generic use. For example, Kleenex is a trademark but came to be used for all brands of facial tissue in common speech; however, no other manufacturer may use the term Kleenex on its product or in advertising. In-line roller skates are commonly called Rollerblades, although Rollerblades refers to only one brand of in-line skates. A small bulldozer-like vehicle used in construction is commonly called a Bobcat, but Bobcat is a trademark for one brand of vehicle. Aspirin, shredded wheat, yo-yo, radar, and scuba are all trademarks that fell into common use. In a recent court case, Hasbro, Inc., maker of Play Doh, is fighting for its trademark against Rose Art, maker of Fun Dough. Rose Art is arguing that Play Doh is now a generic term for modeling clay and that its Fun Dough is not an infringement on the Play Doh trademark.

A number of years ago Xerox initiated an advertising campaign to stave off the common use of its name. Commonly, when one wanted to photocopy a document, it was described as xeroxing, with no regard as to whether the photocopying machine used was a Xerox or other brand. Fearing the loss of their trademark, Xerox Corporation reminded people through TV and magazine ads that Xerox was a trade name and was to be used only when referring to Xerox products.

Keep in mind that this chapter is only a brief introduction to patents, trademarks, copyrights, and trade secrets. Volumes of law as well as dozens of books have been written on the subject. What may seem to be a clear decision in choosing the best type of protection for intellectual property is best made after consultation with an attorney.

Chapter Six

Patents as an Information Source

Many librarians and library users think of patents as information to be searched and used only in situations where a person has an invention and wants to see if it has already been invented. Indeed most patent work in libraries does concern the patent search; however, patents can be used as reference tools to locate information for various purposes.

Since approximately 70,000 patents are granted annually in the United States, it is unlikely that a magazine article or any other public disclosure of a patent will exist except in its form as a registered patent—unless, of course, the patent is controversial or revolutionary. The *New York Times* produces a patent column every Monday that highlights the more interesting patents or issues in patenting. But with 1,300 to 1,500 patents issued every Tuesday, it is impossible to be aware of every new patent. The most difficult aspect of dealing with such a large volume of material is that a patent is often the only form in which a new development is published, so that on-line searches of magazines, newspapers, and other current awareness tools are of little help. The most productive way to use patent information is to scan the *Official Gazette* (OG) every Tuesday in selected patent classifications of interest. A technique for searching the OG was given in chapter 4.

But a question larger than *how* one uses patents as an information source is *why* one would want to. Engineers rely, as do many other professionals, on a handful of standard references and periodicals to be aware of current technology, but in doing so and overlooking patents, most of the new developments in their field are missed. It is a librarian's responsibility to point out the usefulness of patents and to encourage professionals to use them.

Engineers will be used as a representative group to point out the rationale for using patents as an information source. But the reasons given here for searching patent information apply to all professions and are especially valuable for librarians in special libraries.

Pointing Out New Directions in Research

Companies that produce consumer products use the patent literature heavily to see what has already been invented and then try to improve on that invention to market a better product.

For example, a granular chemical that is designed to clear clogged drains is available to consumers. To use this product the chemical is poured into a drain and left for about fifteen minutes while its chemical action dissolves foreign objects, like hair, that are clogging the drain. By examining this patent, a chemist who is aware of the importance of convenience to consumers may realize that although this product works well at what it is supposed to do, it is inconvenient to wait fifteen minutes for the product to work. A better product would be one that could be put into a drain and left to do its work without having to rinse it away after fifteen minutes. There is such a product, a liquid drain cleaner, that is poured into standing water and left until it clears the drain. Clearly, to improve on a patent, a person must be aware of how it works and usually this information is available only in the patent document itself.

Prolific inventors use this one technique to give them direction in new research projects. In the 1980s many homeowners became aware of the attractiveness of vertical blinds in place of drapes for covering large window areas. The vertical blinds could be opened and closed easily and were marketed in a variety of colors and patterns. One inventor, however, by studying the patent on the vertical blinds, became aware of two problems. The first was that, like horizontal venetian blinds, vertical blinds were hard to clean. A homeowner had to clean each vertical slat individually: a major task in semiannual housecleaning, but also a formidable task weekly when each slat had to be vacuumed or dusted. The second problem was that once vertical blinds were purchased, they were similar to drapes in that the homeowner was stuck with the same color or pattern. There was no way to change the appearance of the window coverings without buying a completely new set of blinds.

The inventor solved both problems with a simple improvement on the existing patent. She designed a vertical blind that, instead of having a series of individual slats, had every other slat connected at the bottom with a small rectangle of plastic. This maintained the blinds spacing, but also allowed for the slats to be made of a thin strip of fabric. The fabric was woven through the top of one slat, to the bottom of the same slat, then across the bottom to the next slat, until all the slats were covered with a continual strip of fabric. To clean the blinds or to change colors or patterns, a person needed only to connect another thin strip to the end of the blind and then pull the old fabric strip out. As the strip threaded out, it pulled the new strip through, eliminating the need to thread the new strip by hand, and also making it easy to clean the fabric by throwing it in the washing machine or to change colors and patterns by simple installing a new fabric strip.

Without seeing how the vertical blind worked by studying the patent, the inventor would not have had the inspiration for an improvement on its design. All previous public disclosure of the vertical blinds had been in fashion or decorating magazines that did not describe how the blinds operated.

Many years ago another inventor used the same technique in designing a new type of fishing reel—the spinning reel. It seems backlash, the rolling of the line off the reel after the line had landed in the water, was a major problem. The reel had to spin freely to allow the line to spool in casting, but the reel's tendency to keep rolling caused awful tangles on the line still on the reel. The inventor fiddled with all types of clutches on the reel before studying the patent and realizing that the reel was designed incorrectly.

Any design that allowed the fishing line to feed off the reel in a forward motion would cause backlash. The proper way to feed line was sideways, without the reels moving at all. By studying existing patents the inventor realized that all fishing reels were of this poor design. He designed a sideways reel—the spinning reel—and the problem of backlash was eliminated.

Knowing What Needs to Be Invented, But Not How to Do It

In some cases, engineers in a particular industry know what things need to be invented, but struggle to find a technology that will allow them to do it. Applying an existing technology to a new application is usually the way these types of new inventions come about. The best example of this is video tape.

In the early days of network television, the only way to record a program that was being televised was to point a film camera at the screen and film the broadcast. This left a lot to be desired since the quality was often very poor, and if there were transmission problems with the individual set that was being filmed, the copy was lost.

Engineers struggled for years trying to discover a way electronic television signals could somehow be captured, saved, and then be made to redisplay the image coming over the airwaves. Some ideas were just not economically feasible. The idea was to create a television library housing tens of thousands of broadcasts, and a cheap recording method was necessary.

One television executive, who understood that television worked by making electronic particles bombard a surface in a specific pattern, knew there had to be a way to replicate a simple electronic pattern. This executive, as the story goes, "commanded video tape to be invented." The answer was found by looking to audio tape where sound patterns are recorded magnetically and then reproduced. Video tape does the same thing by recording electronic patterns magnetically on cheap iron oxide plastic tape.

On October 3, 1952, the electronics division of Bing Crosby Enterprises, Inc., in Los Angeles recorded images on magnetic tape, rewound the tape, and immediately reproduced the picture through a standard television monitor. The process was further developed by the Radio Corporation of America (RCA) for the National Broadcasting Company (NBC) and on October 23, 1956, the first video tape program, the Jonathan Winters Show, was broadcast coast to coast by WRCA-TV in New York.

Things that need to be invented are often so different from anything else that exists, that only truly brilliant people can conceptualize them. To understand truly new inventions, people often link them to technologies with which they are familiar. This can be seen in the names given to new technologies. Inventions are often called by names that relate to old technologies.

For example, the automobile was a revolutionary invention but the term automobile was not immediately popular because the public had to get used to a machine that moved without the assistance of a man or animal. Instead, the automobile was initially called a "horseless carriage," something the public could readily visualize. Another example is the airplane. Early passenger aircraft were called "airships" even though an airplane neither behaves nor looks like a ship. The ship metaphor continues today with "spaceships."

New Uses for Existing Technologies

Engineers use the patent literature for still another purpose—to take existing patents and find uses for the invention other than the purpose for which it was invented. The non-stick surface trademarked as Teflon by DuPont was originally used as a surface for pots and pans. But engineers and chemists soon found that a surface that causes very low friction with any other surface that contacts it has a variety of uses.

This information was useful when the medical profession found that artificial heart valves had a tendency to gather plaque from the body since artificial valves could not clean themselves as does most body tissue. This necessitated the removal or cleaning of the valves on a regular basis—a very expensive and dangerous procedure. However, by manufacturing heart valves made of Teflon, the valves opened and closed more easily and did not gather plaque since the low friction prohibited anything from sticking to them. In the same way, the skiing industry discovered that skis covered with Teflon moved more swiftly and smoothly over snow, and a variety of manufacturing industries found uses for Teflon in machines where friction was a problem.

But the most unique application of Teflon was in the garment industry. By studying the structure of Teflon given in the patent, chemists discovered that the molecule was such that it was larger on one side than another. This meant that large molecules such as water could not penetrate a fabric made of Teflon but that smaller air particles could pass through in the other direction. This discovery led to the invention of GoreTex, a weatherproof Teflon fabric used in everything from raincoats to running shoes. GoreTex provides waterproof protection, but unlike plastic or rubber it also lets the fabric breathe, thus allowing the wearer to be more comfortable.

Waiting for Devices to Be Invented

Because engineers scan newly issued patents on a regular basis, they become familiar with what companies and individuals are working on and can sometimes determine the direction of research, especially in a field related to their own. This becomes a valuable tool when an engineer realizes that because of the progress a particular company has made, that company will be able to apply for a patent before him or her.

This is not always a bad situation. At times an engineer, realizing that another company will receive a patent on an important device in a given field, will halt work on the same device. This allows the engineer to save money and resources and to take another direction. At the same time, the engineer's company waits for the patent by the other company to be issued, then contacts the company to license the technology. This happens frequently.

The matter of studying patents has another benefit to an engineer whose company is in competition with another. Since patent indexes list patents by patentee and assignee, it is a simple matter to look up ABC Company and see what patents they have been granted and in what classifications. In other words, it is possible to keep tabs on the competition and to see what they are doing and in which direction they are headed.

Finding What Is Patentable

In the 1980s many biotechnology companies were working in the field of biogenetics, or gene manipulation. Some of these companies had legitimate concerns that their new discoveries would not be allowed protection. Chemical changes in plants, fertilizers, and the like were of no concern, but it did not seem certain that research aimed at genetically manipulating living things would be protected. There has always been a philosophical debate as to whether life could be patented, that is, whether a scientist could "own" a particular variety of, say, a mouse.

A white mouse from Harvard made history in 1988 by becoming the first animal ever to be patented in the United States. Philip Leder and Timothy Stewart of Harvard developed the mouse through genetic engineering by inserting a human cancer gene in mouse egg cells. Because such a creature was novel and useful (two of the primary criteria for a patent) for future cancer research, the U.S. Patent and Trademark Office (PTO) determined that the "Harvard Mouse," as it is now often called, satisfied the requirements for protection as an invention under standard patent law. In making that decision, the patent office removed the last obstacle to patenting any form of non-human life.

Other inventions are granted patent status quietly, although their qualifications to become patents are questionable. The Hula Hoop caused a sensation in the 1950s and quickly became a national craze; however, the idea was not new. The idea of taking a hoop of lightweight material and spinning it around one's waist, keeping it from falling by centrifugal force, had been around for generations. But although the idea was not new, the materials and method used to make the hoop was. The inventor made it from lightweight plastic and connected the ends together by stapling a piece of wood between the ends and molding the inside of the hoop with ridges to give it friction against the body. These were all new developments and improvements and were considered innovative enough to be granted patent protection, but it also made imitation easy in that other companies could produce knock-off hoops simply by changing the connection method and molding different designs into the plastic. Wham-o owned the patent and tried to protect itself by developing new hoops, like the Shoop Shoop Hula Hoop, which had plastic balls inside the tubing allowing for a swooshing sound as the hoop was spun, but the craze died out in little over a year. However, the Hula Hoop was very financially rewarding to Wham-o, which moved ahead with a new invention, the Frisbee.

The Frisbee was another product that had questionable status as a device eligible for patent protection. As the story goes, college students in Boston had discovered that the empty pie plates made by the Frisbee Baking Company had aerodynamic properties when thrown upside down. The students played lunchtime games (after they had eaten the pie) by tossing the pie plates to each other.

An engineer who studied the plates discovered that the shape of the plate was such that the air flowed over the top of the plate faster than under the bottom, giving the plate lift. The weight of the plate kept it stable as it was lifted in flight. Wham-o made a plastic version of the plate, claiming it was new and novel for three reasons: Its center was thinner than its edges, which gave it stability; the top was covered by ridges, which helped give it lift; and the top was curved rather than having a flat surface like a pie plate. Edward Hedrick was granted patent 3,359,678 on a "flying saucer," which he assigned to Wham-o.

Predicting Growth Industries

Engineers use patents to follow the path of "technology transfer." Technology transfer is the process of getting an invention that has been demonstrated successfully in a laboratory into the marketplace. Edison's experience with the electric lamp (see chapter 1) demonstrates this best. Just because an invention works and is patented does not necessarily mean that it is ready for mass consumption or even for use in a limited industry.

Businesspeople and investors use the concept of technology transfer to target the hot stocks for the future. Very basically, using technology to predict growth industries works like this. There is a lag time of ten to twelve years before new developments in a given industry cause that industry to grow. By discovering what patent areas show the most activity today, investors can buy related stocks at low prices and hold them until the new inventions cause economic growth.

Of course, the trick is discovering those growth industries, but that is not as difficult as one might think. A librarian trained in patent use can discover this in two ways. Each year the Office of Technology Assessment (OTA) produces a report called the OTA Forecast (OTAF). In one section of the OTAF, the OTA, a branch of the Commerce Department, lists patent classifications and shows how many patents have been granted in these areas over the past year. Obviously, those areas that are doing the most research are obtaining the most patents. Those are the growth areas for the future. Another method is to scan the OG each week to see which classifications have the most patents granted.

This method, however, is not foolproof. An investor would have to hold onto an investment for a number of years before it showed any significant profit. Also, classifications that show increased activity may not accurately indicate the major industries in ten years. For example, in the early 1970s during the oil embargo, activity in patents relating to synthetic fuels and solar energy increased. But today the synfuels industry is not a major U.S. industry and solar energy, although it has applications in some areas, has not replaced electricity. Recently the classifications showing the most activity are those related to the environment, for example, contaminant cleanup, trash removal, and landfill improvements. Will these be the major industries in 2004?

There are interesting examples of technology transfer and how it was successful given the right amount of time. The IBM Selectric typewriter, patented in the 1970s, allowed a typist to change type fonts by inserting a metal ball element. This was an improvement over the old typewriters that had characters at the end of a series of metal rods. Typewriter fonts could not be changed on these machines. The idea of the ball element, however, was not new. A patent from the early twentieth century was given for a "portable type-writer" which could fit in one's hand. It contained a ball element also.

The story of Chester Carlson and the photocopy machine was explained in chapter 1. The process of electrophotography was patented in 1942, but it was twenty years before the first photocopy machine appeared on the market. This brings up an interesting point. If it takes such a long time to get a newly patented invention into the marketplace, does not the patent expire shortly thereafter? The answer is yes, but good inventors do not get a patent and then sit back and wait for the money to roll in.

Anticipating the long wait for a product to be economically successful, inventors continue to work on the invention, preparing improvements that may be patented. When the product does hit the market, the inventor is prepared with newer versions of the invention. If others compete with similar devices, the improvement is marketed, making the older version obsolete.

This is another reason to use the patent documents. If one is working on a better mousetrap to compete with an existing patented mousetrap, it's good to know if a better mousetrap has already been patented.

Examples of Delays in Technology Transfer

There are many examples of the length of time between an invention's being granted a patent and the time the patented invention makes an impact on the marketplace. Television is an invention that has had a significant impact on society.

Television, rather than being a single invention, is a combination of many inventions relating to both the sending and receiving of signals. The patent that put all the pieces together, however, was the television tube. Once this piece was in place it was technologically possible to send images over the air. But since the technology transfer involved obtaining funds sufficient to build equipment, obtaining the right people to guide the new technology, manufacturing enough affordable receivers to allow the general public to receive the signal, and designing programming to broadcast, it took a long time to begin network broadcasting. Network broadcasting in the United States began in 1948. The picture tube was invented by Vladimir Zworykin and was granted patent 2,139,296 in 1929. In this case, technology transfer took nineteen years.

Even older patents had a lag. The technology transfer involved in the inventions of the electric lamp and the telephone were mentioned in chapter 1, but these were fairly complicated devices. A simpler device, but one having no less impact on society, was the invention of the mechanical reaper by Cyrus McCormick in 1834. McCormick obtained patent 8,277 for his reaper, and waited until 1850 before he was able to secure the backing necessary to market his invention.

Advantages of Patents as a Research Tool

In summary, librarians and patrons should become familiar with the techniques of using patents as an information source. The many advantages of being able to do this include:

Seeing what has been invented

Checking on what the competition is doing

Waiting for desired products to be granted patents

Finding the most up-to-date information, patents often being the only published form in which a new idea appears

Being aware of technology transfer

Pointing out new directions in research

Improving existing inventions

Knowing what needs to be invented

Finding new uses for existing patents

Finding what is patentable

Targeting growth industries of the future

For librarians, knowing how to use patents as an information source enhances reference services. Much of the information found in patents is not available elsewhere. For those librarians assisting people in business and industry, patent savvy is a must. We hope we've provided some of that savvy here.

Bibliography

General Information Concerning Trademarks. Washington, D.C.: Govt. Printing Office, 1982.

Guide for Filing Design Patent Application. Washington, D.C.: U.S. Department of Commerce, 1989.

Ardis, Susan B. *An Introduction to U.S. Patent Searching.* Englewood, Colo.: Libraries Unlimited, 1991.

Auger, C. P., ed. *Information Sources in Patents.* London: Bowker-Saur, 1992.

Beggs, John J. *Long Run Trends in Patenting.* Cambridge, Mass.: National Bureau of Economic Research, 1982.

Carr, Fred K. *Searching Patent Documents for Patentability and Information.* Chapel Hill, N.C.: F. K. Carr, distributed by Patent Information, Inc., 1982.

Foltz, Ramon D. and Thomas A. Penn. *Protecting Scientific Ideas and Inventions.* Boca Raton, Fla.: CRC Pr., 1990.

Foster, Frank H. and Robert L. Shook. *Patents, Copyrights and Trademarks.* New York: Wiley, 1989.

Griffin, Gordon D. *How to Be a Successful Inventor.* New York: Wiley, 1991.

Griliches, Zvi. *The Value of Patents as Indicators of Inventive Activity.* Cambridge, Mass.: National Bureau of Economic Research, 1986.

Jazwin, Richard. *Inventing from Start to Finish.* Troy, Mich.: Business News Publishing Co., 1991.

Kivenson, Gilbert. *The Art and Science of Inventing.* New York: Van Nostrand, 1977.

Levy, Richard C. *Inventing and Patenting Sourcebook.* Detroit: Gale, 1992.

Petroski, Henry. *The Evolution of Useful Things.* New York: Knopf, 1992.

Redmon, Tina. *The Inventor's Handbook on Patent Applications.* New York: Vantage, 1993.

Electronic Services

More than thirty different online files are offered by various vendors. In addition, patent files and indexes are available on CD-ROM and through the Internet. Following are some of the more popular automated patent services.

Service	File
Dialog	Claims
Dialog	Derwent World Patents Index
LEXIS/NEXIS	Lexpat
Micropatent	20-year backfile on CD-ROM
Mosaic site	http://sunsite.unc.edu/patents/intropat.html
	500 MB of patent information

It should be noted, however, that electronic databases are not fully retrospective, often going back only to the 1970s. Drawings are provided only on CD-ROM products, at this writing. Purchase fees for CD-ROM and online usage fees should be considered in deciding the search method to be used. A "quick and dirty" electronic search can save much time in certain circumstances. The manual method assures comprehensiveness, drawings, and low cost.

Thirty Often Asked Questions and Their Answers for the Amateur Inventor

1. You invent a widget. Your brother gives you $1,000 to finance the manufacture of the widgets, with the understanding that you each get 50 percent of the profits on sales of the widgets. To ensure this, you file the patent application in both your names, right?

   A. WRONG. Patent applications must be filed in the name of the true inventor(s). If noninventors are named in the application, it can invalidate the patent.

2. After a reasonable patent search, you file an application for a patent. You find out shortly after mailing your application to the PTO that the invention already exists. Are you liable as an infringer?

   A. NO. An application is not an infringement. Only the manufacture, sale, or use is infringement.

3. Are advertising slogans, written statements on T-shirts, or bumper sticker sayings such as "Where's the Beef?" protected by copyright?

   A. NO. Copyright is reserved for prose, poetry, maps, artistic works, computer programs, and movies. Advertising slogans are trademarks.

4. Your brother invents a widget, but is too lazy to patent it. Can he transfer his rights to you and let you patent it?

   A. NO. See the answer to question 1.

5. By filing a Document Disclosure with the PTO, an inventor has a two-year period to file an application without losing the rights to an invention, right?

   A. WRONG. The only thing a Disclosure will do for you is show evidence of the date of inception of an idea and provide a reliable witness.

6. Isn't it true that filing an application is not recommended because others can see, copy, or steal your invention by going to the PTO and looking at the filed applications?

   A. NO. Applications are filed in secrecy by law. Outsiders have no access to patent information.

7. Twenty years ago you invented something but never applied for a patent. You find out that somebody else has recently patented your invention. Do you have any rights to the invention, even if you can prove the date of your invention?

   A. NO. The delay in patenting your invention negates your rights.

8. If an invention has been patented by somebody else and the patent has expired, can someone else get a patent on it?

   A. NO. You cannot get a patent on anything that has been previously patented. Also, you cannot get a patent on anything shown in a magazine or any other publication anywhere in the world whether or not it is patented. Legally, this is called "anticipation."

9. If you see an invention that is patented in France, can you get the U.S. patent on it?

   A. NO. Only the true inventor has the right to patent in other countries. Also, the publication of a patent in France would come under anticipation laws.

10. OK. So you can't patent the French invention, but you decide to go ahead and manufacture it in the United States anyway. Can the French inventor stop you?

    A. NO. A patent is enforceable only within the country in which it was issued. He may send Big Pierre to break your thumbs, but legally he can't touch you.

11. Inventors often mail themselves a drawing and description of their invention by certified mail. Is this a good way to protect your rights as an inventor?

    A. NO. Certified mailers are poor substitutes for a signed disclosure by a witness. Amazingly enough, some patent attorneys still recommend this method of protection for their clients.

12. Something you have invented appears in the drawings of somebody else's patent, but no mention is made of it in the claims. If you manufacture your own product, are you liable?

    A. NO. Only the claims of a patent determine infringement.

13. Do you have to wait until your patent is granted before you can sell or license your invention to business?

A. NO. Corporate buyers want to purchase inventions as soon as possible after an application is made so that they can get a jump on the competition before an invention is made public.

14. Does patent pending mean that you can sue people who are copying and manufacturing your invention?

    A. NO. Patent pending confers no legal rights; however, after the patent is granted, you may sue an infringer.

15. Do you have to prepare a working model of your invention for submission to the PTO before your patent will be granted?

    A. NO. The PTO has not required models for many years.

16. After seventeen years a patent can be renewed for another seventeen years, right?

    A. WRONG. It takes an act of Congress to renew a patent. It has never been done.

17. If a device isn't marked "patented" and doesn't have a patent number, can you copy it?

    A. NO. Patents don't have to be marked.

18. Can a patent be declared invalid or taken away from you?

    A. YES. It can be declared invalid for many reasons, including overlooked anticipation.

19. If somebody infringes on your patent, and you file a report with the PTO, will they take action against the infringer?

    A. NO. The PTO is not a law enforcement agency. You alone are responsible for protecting your rights.

20. If you have an invention that you think would interest a company, and you send a description of it to them, will the firm agree to keep the information confidential and pay you if it uses the product?

    A. NO. The company will not agree to these terms. Instead, it will require you to sign one of its forms stating that the company has no obligation to you regarding payment or confidentiality.

21. Is it true that words like kleenex, thermos, radar, and sonar were once trademarks, but are now generic terms allowing anybody to use them on goods?

    A. NO. These terms are still valid trademarks; however, allowing a trademark to fall into generic use makes legal protection difficult. Xerox went through an extensive campaign several years ago to remind people that Xerox was not a generic term.

22. Sometimes instead of seeing the standard R in a circle which designates a registered trademark, you will see a T in a circle. Do these symbols mean the same thing and are they interchangeable?

    A. NO. Trademarks have to be used in interstate commerce before they can be registered. The circled T symbolizes that it is being used as a trademark, but is not yet registered.

23. If you change one or two letters of a trademark in use, for example, changing xerox to xeroks, can you use it on your goods without infringing on an existing trademark?

    A. NO. A trademark has to be different enough to avoid confusion, mistake, or deception. Usually changing one or two letters is not enough.

24. Do you have to register your trademark before using it on your goods?

    A. NO. Trademarks have to be used *before* they are registered.

25. Copyright can be used to protect ideas, systems, or methods of doing things, right?

    A. WRONG. Copyright can be issued only on a *form* of expression, not the idea per se.

26. Is it true that a trademark is a graphic symbol while a trade name is a word?

    A. NO. A trademark is any word, symbol, or sound used to identify goods.

27. Before a copyright is issued, the Copyright Office goes through an extensive search, in much the same way as the PTO does for patents and trademarks, right?

    A. WRONG. The Copyright Office doesn't concern itself with whether or not a work is previously copyrighted. If there is an infringement on copyright, it is up to the individual and the courts to straighten things out.

28. If you make a cassette tape from an album you have at home so that you can play the music in your car, can you be prosecuted for violation of copyright law?

    A. YES. Although rarely enforced, this is a violation of copyright.

29. You invent an electric fork and attempt to trademark it with the name "Electric Fork." Can you do this?

    A. NO. A product's common name, even if its function is unique, cannot be protected.

30. Can sounds be copyrighted?

    A. YES. Although very few are, the MGM lion's roar and the NBC chime are two examples.

Patent and Trademark Depository Libraries in the United States

Science and Technology Department
Ralph Brown Draughon Library
Auburn University, AL 36849-5606
Phone: (205) 844-1747

Government Documents Department
Birmingham Public Library
2100 Park Place
Birmingham, AL 35203
Phone: (205) 226-3680

Reference Services, Anchorage
Municipal Libraries
Z. J. Loussac Public Library
3600 Denali Street
Anchorage, AK 99503-6093
Phone: (907) 261-2916

Daniel E. Noble Science and
Engineering Library
Arizona State University
Tempe, AZ 85287
Phone: (602) 965-7607

State Library Services
Arkansas State Library
One Capitol Mall
Little Rock, AR 72201-1081
Phone: (501) 682-2053

Science, Technology and Patents
Los Angeles Public Library
630 West Fifth Street
Los Angeles, CA 90071-2097
Phone: (213) 612-3273

California State Library
Library-Courts Building
914 Capitol Mall
Sacramento, CA 94237-0001
Phone: (916) 322-4572

Science and Industry Section
San Diego Public Library
820 E. Street
San Diego, CA 92101
Phone: (619) 236-5813

Sunnyvale Patent Clearinghouse
Sunnyvale Public Library
1500 Partridge Avenue, Bldg. 7
Sunnyvale, CA 94087
Phone: (408) 730-7290

Business, Science and Government
Publications Department
Denver Public Library
1357 Broadway
Denver, CO 80203
Phone: (303) 640-8847

Science Park Patent/Technology Library
25 Science Park
New Haven, CT 06511
Phone: (203) 786-5447

Reference Department
University of Delaware Library
Newark, DE 19717-5267
Phone: (302) 451-2965

Undergraduate Library
500 Howard Place, N.W.
Howard University
Washington, DC 20059
Phone: (202) 806-7252

Government Documents Department
Broward County Main Library
100 South Andrews Avenue
Fort Lauderdale, FL 33301
Phone: (305) 357-7444

Business and Science Department
Miami-Dade Public Library
101 West Flagler Street
Miami, FL 33130-2585
Phone: (305) 375-2665

Reference Department, Library
P.O. Box 25000
University of Central Florida
Orlando, FL 32816-0666
Phone: (407) 823-2562

University of South Florida
Patent Library
Library LIB 122
Tampa, FL 33620-5400
Phone: (813) 974-2726

Department of Microforms
Price Gilbert Memorial Library
Georgia Institute of Technology
Atlanta, GA 30332-0900
Phone: (404) 894-4508

Federal Documents Department
Hawaii State Public Library System
478 South King Street
Honolulu, HI 96813
Phone: (808) 586-3477

Library
University of Idaho
Moscow, ID 83843-4198
Phone: (208) 885-6235

Business/Science/Technology Division
Chicago Public Library
Harold Washington Library Center
400 South State Street
Chicago, IL 60605
Phone: (312) 747-4400

Reference Department
Illinois State Library
Centennial Building
Springfield, IL 62701-1796
Phone: (217) 782-5659

Business, Science and Technology
    Division
Indianapolis-Marion County Public
    Library
P.O. Box 211
Indianapolis, IN 46206
Phone: (317) 269-1741

Siegesmund Engineering Library
Potter Center
Purdue University
West Lafayette, IN 47907

Head, Information Services & Patent
    Depository
State Library of Iowa
East 12th & Grand
Des Moines, IA 50319
Phone: (515) 281-4118

Ablah Library
Wichita State University
Wichita, KS 67208-1595
Phone: (316) 689-3155

Reference and Adult Services
Louisville Free Public Library
301 York Street
Louisville, KY 40203-2257
Phone: (502) 561-8617

Business Administration/Government
  Documents Department
Troy H. Middleton Library
Louisiana State University
Baton Rouge, LA 70803
Phone: (504) 388-2570

Reference Services
Engineering and Physical Sciences
  Library
University of Maryland
College Park, MD 20742
Phone: (301) 405-9157

Physical Sciences Library
Graduate Research Center
University of Massachusetts
Amherst, MA 01003

Boston Public Library
Copley Square
P.O. Box 286
Boston, MA 02117
Phone: (617) 536-5400

Engineering Library
3074 B. H. H. Dow Building
University of Michigan
Ann Arbor, MI 48109-2136
Phone: (313) 764-7494

Ferris State University
Abigail S. Timme Library
901 South State Street
Big Rapids, MI 49307

Technology and Science Department
Detroit Public Library
5201 Woodward Avenue
Detroit, MI 48202
Phone: (313) 833-1450

Technology and Science Department
Minneapolis Public Library
300 Nicollet Mall
Minneapolis, MN 55401
Phone: (612) 372-6570

Mississippi Library Commission
1221 Ellis Avenue
P.O. Box 10700
Jackson, MS 39289-0700

Reference Department
Linda Hall Library
5109 Cherry Street
Kansas City, MO 64110
Phone: (816) 363-4600

Applied Science Unit
St. Louis Public Library
1301 Olive Street
St. Louis, MO 63103
Phone: (314) 214-2288

Patent Center
Montana College of Mineral Science
  and Technology Library
Butte, MT 59701
Phone: (406) 496-4281

Engineering Library
Nebraska Hall, 2nd Floor West
University of Nebraska-Lincoln
Lincoln, NE 68588-0410
Phone: (402) 472-3411

Government Publications Department
Getchell Library
University of Nevada-Reno
Reno, NV 89557-0044
Phone: (702) 784-6579

Patent Collection
University Library
University of New Hampshire
Durham, NH 03824
Phone: (603) 862-1777

Newark Public Library
P.O. Box 630
5 Washington Street
Newark, NJ 07101-3175
Phone: (201) 733-7782

Government Documents Department
Library of Science and Medicine
Rutgers University
Piscataway, NJ 08855-1029
Phone: (908) 932-2895

Centennial Science and Engineering
   Library
University of New Mexico
Albuquerque, NM 87131
Phone: (505) 277-4412

Reference Services, New York State
   Library
Cultural Education Center
Empire State Plaza
Albany, NY 12230
Phone: (518) 473-4636

Science and Technology Department
Buffalo and Erie County Public Library
Lafayette Square
Buffalo, NY 14203
Phone: (716) 858-7101

Patents Collection
New York Public Library at 43rd Street
521 West 43rd Street
New York, NY 10036-4396
Phone: (212) 714-8529

D. H. Hill Library
North Carolina State University
P.O. Box 7111
Raleigh, NC 27695-7111
Phone: (919) 515-3280

Chester Fritz Library
University Station
University of North Dakota
Grand Forks, ND 58202
Phone: (701) 777-4888

Science and Technology Department
Public Library of Cincinnati and
   Hamilton County
800 Vine Street
Cincinnati, OH 45202-2071
Phone: (513) 369-6936

Documents Collection
Cleveland Public Library
325 Superior Avenue
Cleveland, OH 44114-1271
Phone: (216) 623-2870

Information Services Department
Ohio State University Libraries
1858 Neil Avenue Mall
Columbus, OH 43210
Phone: (614) 292-6175

Science/Technology Department
Toledo/Lucas County Public Library
325 Michigan Street
Toledo, OH 43624
Phone: (419) 259-5212

Oklahoma State University Library
Stillwater, OK 74078-0375
Phone: (405) 744-7086

Documents Section
Oregon State Library
State Library Building
Salem, OR 97310
Phone: (503) 378-4239

Documents Department
The Free Library of Philadelphia
Logan Square
Philadelphia, PA 19103
Phone: (215) 686-5331

Science & Technology Department
Carnegie Library of Pittsburgh
4400 Forbes Avenue
Pittsburgh, PA 15213
Phone: (412) 622-3138

Documents Section
C207 Pattee Library
Pennsylvania State University
University Park, PA 16802
Phone: (814) 865-4861

Business/Industry/Science/Patent
    Department
Providence Public Library
225 Washington Street
Providence, RI 02903
Phone: (401) 455-8027

Library
Medical University of South Carolina
171 Ashley Avenue
Charleston, SC 29425
Phone: (803) 792-2372

R. M. Cooper Library
Clemson University
Clemson, SC 29634-3001

Memphis & Shelby County Public
    Library and Information Center
1850 Peabody Avenue
Memphis, TN 38104
Phone: (901) 725-8876

Vanderbilt University
Stevenson Science Library
419 21st Avenue South
Nashville, TN 37240-0007
Phone: (615) 322-2775

McKinney Engineering Library
Room 1.3 ECJ
University of Texas at Austin
Austin, TX 78712
Phone: (512) 471-1610

Sterling C. Evans Library
Room 217
Texas A&M University
College Station, TX 77843-5000
Phone: (409) 845-2551

Government Publications Division
Dallas Public Library
1515 Young Street
Dallas, TX 75201
Phone: (214) 670-1468

Division of Government Publications &
    Special Resources
Fondren Library
Rice University
Box 1892
Houston, TX 77251-1892
Phone: (713) 527-8101

Documents Division
Marriott Library
University of Utah
Salt Lake City, UT 84112
Phone: (801) 581-8394

Documents and Interlibrary Loan
    Services
Virginia Commonwealth University
P.O. Box 2033-901 Park Avenue
Richmond, VA 23284-2033
Phone: (804) 367-1104

Engineering Library, FH-15
University of Washington
Engineering Library Building, FH-15
Seattle, WA 98195
Phone: (206) 543-0740

Evansdale Library
West Virginia University
P.O. Box 6105
Morgantown, WV 26506-6105

Technical Reports Center
Kurt F. Wendt Library
University of Wisconsin-Madison
215 North Randall Avenue
Madison, WI 53706
Phone: (608) 262-6845

Milwaukee Public Library
814 West Wisconsin Avenue
Milwaukee, WI 53233
Phone: (414) 278-3247

Patent Documents in Federal Depository Libraries

The classification numbers at the end of each entry are SuDocs numbers. This classification scheme is used to catalog government documents. Ask for assistance in locating material by this classification number.

Attorneys and Agents Registered to Practice before the United States Patent and Trademark Office.
C 21.9/2:year

Classification Definitions (Patents).
C 21.3/2:class number

Concordance: United States Patent Classification to International Patent Classification.
C 21.14/2:C 74/year

Design Patents.
C 21.24:D 46/year

General Information Concerning Patents: A Brief Introduction to Patent Matters.
C 21.26/2:year

General Information Concerning Trademarks.
C 21.26:year

Guide for the Patent Draftsmen: Selected Rules of Practice Relating to Patent Drawings.
C 21.14/2:D 78/980

Index of Patents Issued from the United States Patent and Trademark Office.
- List of Patentees   C 21.5/2:year/pt. 1
- Index to Subjects of Inventions   C 21.5/2:year/pt. 2

Index of Trademarks Issued from the United States Patent and Trademark Office.
C 21.5/3:year

Index to the United States Patent Classification.
C 21.12/2:year

Manual of Classification of Patents.
C 21.12/2:982

Manual of Patent Examining Procedure.
C 21.15/:year

Official Gazette of the United States Patent and Trademark Office:
Patents.
C 21.5:volume/number

Official Gazette of the United States Patent and Trademark Office:
Trademarks.
C 21.5/4:volume/number

Patents and Inventions: An Information Aid for Inventors.
C 21.2:P 27/10/977

Index

Timothy Lee Wherry is head librarian at the Robert E. Eiche Library at the Pennsylvania State Altoona campus. He has been writing and speaking on the subject of patents for thirteen years and has worked as an engineering librarian at Carnegie-Mellon University, Arizona State University, and the Phoenix (Ariz.) Public Library.